The Car Opening Business

A Manual for
Operating Your Own
Lockout Service
Business

Dennis Collins
Steve Cormier

QUICK START LOCKSMITH TRAINING

BUDGET LOCK & KEY INC.
6547 N. Academy, Ste. 532
Colorado Springs, CO 80918
719-598-2135
QUICK START LOCKSMITH TRAINING

Warning-Disclaimer

This manual is designed to give you information regarding car opening as it relates to the locksmithing industry and to the subject matter covered herein. It is sold with the understanding that the author(s) or publisher(s) are not engaged in rendering legal, accounting, or other professional advice. If legal advice or other professional assistance is required, the services of a competent professional person and/or company should be sought.

It is **not** the purpose of this manual to provide all of the information on the subjects covered in this manual but to emphasize the highlights of many areas. The purpose of this manual **is** to give you a sound background in the field of car opening as it pertains to the locksmithing industry. What we have tried to do is complement and supplement other materials about locksmithing. It is recommended that you read all available materials about the subject and learn as much as possible about this industry.

Locksmithing is not a get-rich-quick-scheme. Anyone who is devoted to starting a career in this industry will need to invest time and effort. For many people the locksmithing trade has helped them build solid, growing, and very rewarding businesses.

Every effort has been made to make this manual as complete and as accurate as possible. However, there may be mistakes both typographical and in content. Therefore, this manual should be used only as a general guide and not as an ultimate source of locksmithing information. Furthermore, this manual may contain dated information that is no longer accurate.

The author(s) and publisher(s) of this manual are writing this to educate and entertain the readers. Therefore, the author(s) and/or publisher(s) shall have neither liability nor responsibility to any and all persons or entities with respect to any loss or damage caused directly or indirectly by the information contained in this book.

Credits and Acknowledgments

Book design and production: Graphics West, Inc., Colorado Springs, Colorado

Section Art: Bill Crowley, Colorado Smiles
2501 W. Colorado Ave.
 Colorado Springs, CO 80904
(719) 471-2704, Fax (719) 471-7135

Copyright

ISBN 09669082-1-X

6547 N. Academy, Ste. 532
Colorado Springs, CO 80918
719-598-2135

QUICK START LOCKSMITH TRAINING

Contents

Introduction

Purpose of This Manual

The authors, two individual locksmiths each owning his own successful locksmith business, collaborated in writing this manual to bring you, the new car opening entrepreneur, over twenty-five years of combined knowledge and experience in the car opening business. This information is not available anywhere else.

The purpose of this manual is not to give you specific techniques on how to open cars. We do not teach you how to open cars in this manual. Opening cars is easy, you will become a car opening expert very quickly. It would be redundant for us to teach you how to open cars when all of the car opening techniques you need to learn are explained and illustrated in the manual that come with the car opening tool set you will purchase separately.

The hardest part of this business is learning the business. Knowing what works and what doesn't in the areas of advertising and communications can make the difference between success and failure in this business. That is why we have written this manual for you.

Car Opening Expert's Character

A car opening expert is an exceedingly honest man. He has good habits and a keen sense of right and wrong. He cooperates with the police at all times, and never opens a car if he has any doubt about the honesty of the customer. The car opening expert is honorable with a good reputation. He must preserve the confidence and trust that has been built up over many years. He must keep himself above suspicion at all times. The public has a high opinion of the car opening expert so he must always act with integrity and use good judgment.

DENNIS COLLINS   I would like to congratulate you on your exciting new lockout service business. In my opinion the lockout service business is one of the easiest businesses to own and operate. You can make your total investment back very quickly by opening just a few cars.

Before we begin, I would like to tell you a little something I learned about being successful. In order to become successful you must be willing to start setting goals for yourself. If you are not willing to take the necessary time and the effort to reach your goals and realize your dreams ... why bother?

However, if you are excited about the potential of making a good income and working for yourself, this manual is for you. We will help you reach those goals and realize those dreams. This is a business you can build as big as you want it to be!

A little about myself, I have worked for several different locksmith companies over the last sixteen years. Most locksmith companies are mom-and-pop operations so I worked very closely with the owners and was involved in every aspect of their businesses. The first job I had with a locksmith company was the hardest job I ever had in my life. I worked seven days a week, twenty-four hours a day for a salary of $175 a week. I received no commissions, no benefits, no overtime pay, or anything extra and I did this for three years. I kept this job because I loved the trade and we lived in such a small town that I believed this was my only employment opportunity in locksmithing. I quit when I finally burned out. Looking back on it now, I can see that in three years I gained the equivalent of almost ten years of locksmithing experience and more practical business experience than I could have ever learned in school. I was surprised at how easy it is to get another locksmithing position if you have some experience. This is true even today; there is a shortage of locksmiths. I also discovered that most of these companies are learning the trade as they go, so don't let the fact that you are a beginner bother you. Most locksmiths, though they will not admit it, feel like beginners as well.

I want to emphasize to you that hard work alone will not make you successful in this business. You must know how to "work smart" and that is exactly what we will teach you.

I became successful because I was determined to make it happen. I had come to a point in my life where I just could not stand to work for anyone else any more! I thought of all the money I had made for my employers, and I was having trouble paying the rent on my small apartment! Just four years after I started my own business I moved out of that small apartment and into a brand new two-story house that I had custom built. My service van is a full-size Chevrolet—right off the showroom floor. That is a huge jump in lifestyle in only four years!

You must begin setting goals. My first goal in this business was to set the date that I would become self-employed. I had no money and no one to help, but I was determined to do this and I found a way. You can too. It takes a lot of "smart work" and the hours can be long at times, but I wouldn't have it any other way.

The idea of writing this manual came to me after years of hearing my customers saying to me, "$45 for that? I should be in your business."I heard this so many times that I began wondering why there weren't more of us doing this. The answer is simple. If you do not know the right approach, this business can be almost impossible to get into. Suppliers will not sell to you, people in the car opening business will not talk to you, and everyone will tell you that you can't do this.

Locksmithing has always been a very secretive trade … until now. We all have the same rights, privileges, and opportunities. Anyone can be a locksmith and I'm going to show you how to make a good living doing it. You may not agree with everything presented to you here. That's fine. If you have other ideas, try them; that's how you learn. I know these methods work very well for me, and they will work for you too.

I would like to say thank you for purchasing this manual and wish you good luck.

STEVE CORMIER I became a car opening expert from a slightly different background. I was a lot like most people that I knew. I had a dead-end job and going no place very fast. I had set a goal to become self-employed at some time in my life, but things did not happen quite as fast as I thought. Here is how I started in the locksmithing and car opening field.

In the early '70s I jumped from job to job trying to find some type of work that was satisfying and paid well. Money was important but if the job was too boring money did not seem to matter as much. Every time I found a satisfying job, the pay would not support my needs.

Then things started to change for me about ten years ago. I started with a company that sold hardware to all the local hardware stores and locksmith shops in the state of Colorado, and one of the product lines was keys. Over time I became acquainted with many of the locksmiths in my area by going into their shops to sell them keys. This, in itself, was not the miracle cure for the monotonous jobs. What it did do was show me the car opening side of locksmithing. Those secrets are what we are sharing with you in this book. Being persistent, I finally gathered enough information to become more and more interested in this great opportunity of locksmithing and car opening. I decided to get more familiar with the idea of working as a locksmith and ordered a correspondence course through the mail. It took two months to get the courage to quit my regular job, although there was a little outside encouragement—money, free time, and no boss. Downsizing of the company that I worked for also helped me make my decision. I should thank that company. With the company in the slumps I needed a new job, and this gave me the opportunity to become self-employed.

I kept seeing locksmiths making lots of money and doing very little to get it. This is how I came to the realization that you can make a good living opening cars. You do not have to know everything or be a top notch locksmith to open cars and make a very comfortable living.

Keep in mind that I followed the steps in this manual even though they were not in writing. The different sections of this book come from the experiences I have every day. I learned each and every part the hard way.

Are you wondering if I am a locksmith? The answer is yes, I am and will be for a long time to come. I enjoy being a car opening expert and helping other people. This has become second nature to me.

⚡ It's A Fact!

This is the best job I've ever had.

This is the best job I've ever had and that's one reason why you should consider this opportunity very seriously. Enjoy the money, the extra time, and the freedom. I was forced into this career change by circumstances; but if I had had a manual like this one, the choice would have been a simple one from the very start.

Step by Step

SECTION 1

Step by Step

Introduction

This section will walk you through the entire manual in a step-by-step sequence to make it very easy for you to put all of the information to work for you very quickly.

We have put together in this manual all the information that you will need to build a successful car opening business. You may already have ideas of your own that you believe would be good for your car opening business but chances are that we have already been there and done that! We have over twenty-five years of experience in the car opening business and will try to make you aware of all the options available to you. We strongly suggest that you follow our recommendations first, and then experiment with other ideas after you become successful.

There are many small car opening businesses currently operating that are barely making enough money to pay their bills. These companies come and go every year. It is almost fun to look in the yellow pages each year just to see who is new and who is out of business. On the other side of the coin, there are some very successful car opening businesses out there, and some of them are even franchising their businesses nationwide. Why are some very successful while others are failing? We have interviewed some of the companies that are barely surviving (which we consider failing). We have discovered that in each and every case the reason that they are failing (and the only reason they are failing) is that they do not have the information that you have in your hands right now!

 **Step 1** ## Decide What You Are Going to Call Yourself

The very first thing that you need to do is to get into a position that will allow you to legally purchase and carry car opening tools. Car opening tools are not sold to just anyone and some cities have possession laws making them illegal unless you are in one of the accepted trades or businesses that are permitted to have them. You will need to meet the criteria that places you in an accepted trade or business.

We will do this by first deciding just what you want to be known as. We're not talking about a business name, but more along the lines of what trade or type of business you are in. Since all you do is open cars, you could call yourself many different things. We strongly suggest that you call yourself a locksmith right from the start and here's why. The locksmithing industry is very hard to get into if you're not already a locksmith. Sounds funny, doesn't it? Here's what we mean. Have you ever heard this before?

"I'm sorry we cannot hire you because you do not have any experience." It's a catch-22; you cannot get a job without experience, and you cannot get experience without a job! An example of this would be: Call a company that sells car opening tools and say, "I'm not a locksmith but I would like to be one. Can I buy a set of your best car opening tools?" They will say, "No, I'm sorry; we can only sell those tools to locksmiths."

So how are you going to be a locksmith if they will not sell you the tools? There are other types of businesses that are allowed to purchase these tools, although they are not as readily accepted into other areas like security and training. Some of them are:

- Tow truck drivers
- Repossessors
- Car dealers
- Police
- Firefighters

Some suppliers do have a "We only sell to locksmiths" policy. So, to take the path of least resistance, we have written this Manual with the assumption that you are going to become a locksmith.

This holds true for the other tools you will need such as lock picks, lock-picking books, and videos. All of these tools and information are considered by the locksmith industry to be secrets available only to the privileged few who call themselves locksmiths. You will, of course, be buying the majority of your tools and books from suppliers who service the locksmith industry. Therefore, you will need to call yourself a locksmith.

It is very difficult to get the tools you will need to run your car opening business if you are not a locksmith. More likely than not, once you get into the car opening business you are going to have so much fun that you are probably going to want to learn how to do more of this type of work. If you are not a locksmith we are back to the catch-22 again.

It will also be easier and cheaper to have your supplies delivered to you if you are a locksmith. According to the United States Postal Service, all locksmith devices mailed within the United States are not allowed to be mailed through the Postal Service unless addressed to one of the following:

- Bona fide locksmith
- Bona fide repossessor
- Lock manufacturer or dealer
- Motor vehicle manufacturer or dealer

If you do not fall into one of the categories listed above, your supplies must be shipped via United Parcel Service (UPS) or Federal Express. Unfortunately UPS is more expensive and takes longer to deliver than the United States Postal Service.

One of the easiest ways to legally order your tools and have these tools in your possession is to become a locksmith. So how do you become a locksmith? All you have to do is start your own locksmith business!

You are probably thinking that this is going to be frustrating and complicated but it's not. We are going to walk you through the five easy steps to starting your own locksmith business right now! You have already completed the first step just by deciding to become a locksmith. Many people have misconceptions about the locksmith industry and these misconceptions help keep the industry's image clean and give your customers a sense of trust in you. These are some of the questions that people ask us about becoming a locksmith:

Do you have to be bonded?

Do you have to be insured?

Do you have to be certified?

Do you have to go to school to learn this?

Are your fingerprints on file at the police department?

Do you have to get a security clearance to do this work?

Do you have to take a lie detector test to be a locksmith?

Do you have to go through an extensive background check?

Do you have to have a spotless police record to be a locksmith?

The honest answer to **all** of these questions is no. If you were to ask us these questions on the street or over the phone we would answer yes to all of them and so would every other locksmith that we know! This discourages people with police records from wanting to become locksmiths because they don't think they will pass these rigorous standards that they believe exist. It also builds an image of honor and trust in the industry.

The truth is that, at this time, the locksmith industry is virtually unregulated. This means there are very few cities or states that require any kind of licenses, permits, or have any other requirements to be met. This is a free market economy and anyone who wishes to become a locksmith can be one by simply hanging out a shingle. (Hanging out a shingle is a slang term for starting a business.)

This is how you start the business. First, decide what you are going to call yourself (we are going to be locksmiths).

Step 2 See If a License or Permit Is Required

The next thing you need to do is to check on whether or not the state you live in requires a license or permit to do the type of work you will be doing. Simply call information and ask for the phone number of the state attorney

⟲ Reminder

Step 1 Decide what you are going to call yourself

Step 2 See if a license or permit is required

general's office. When someone answers tell them you are requesting information on the rules and regulations of operating a locksmith business in your state. Most of the time you will get a recording requesting your name, phone number, and nature of your call. Someone will return your call. Be patient; sometimes it takes a couple of days for a representative to get back to you. If a license or permit is required, you will be notified. You will receive a rules and regulations information package and you will be given the information you need to apply for any license or permit. You will also need to check with the city you live in to see if a license or permit is required to perform locksmith work in that city. This is done by calling your local district attorney's office and requesting their assistance.

We are not talking about a business license or permit here. Right now we are checking to see if a license or permit is required to perform the type of work you will be doing. Any business that you start may require that you have a business license. When you call about rules and regulations you will be told at that time if one is required. If no one mentions it, ask! There is nothing to worry about. Everyone who applies for a business license gets one; you cannot be turned down.

Step 3   Choose a Name for Your Business

Now it's time to choose a name for your business! Please take our advice on this subject. This, in and of itself, can be an asset or a liability to your company. Assets are good for you, liabilities are not. We have seen many companies that we believe could have been successful, but they had such silly names that people were simply afraid to call them for service.

Choose your company name carefully and seriously. You are going to be a mobile service business and the first and possibly the only impression your customers will have of your business will be your company name. People who are locked out of their vehicles are not looking for a shop to go to. They are locked out of their cars and are looking for someone to come to them. They will begin to form an opinion of your company simply from the name you've chosen. This **can** affect the number of calls you get. Your business name can make you sound like a large corporation or a small mom-and-pop operation. Your customers will never know the difference as long as you perform your service in a professional manner!

We can tell you what makes a name work. **Familiarity.** People tend to choose service providers because they feel like they have either heard of, or seen them before. People are looking for something they can relate to.

Everyone seems to be afraid of looking foolish by making the wrong decision. An excellent example of this: Imagine this scene. We are working on one of fifteen storefront doors. We mean fifteen doors side-by-side at a large superstore. Ninety-five percent of the people coming into and going out of the store will use the door that we are working on! Now, are they doing this just to disturb our work and make our job take ten times as long? No! They do this because they can see that we are working on one door and they know that it is not locked. How foolish would they look if someone saw them try to open one of the other doors and find it was locked?

We are trying to explain to you what we call a common obstacle to business success. A large part of your public acceptance in the service industry comes from the impression or image that a customer perceives from your business name.

This is one reason people will refer your company to others. If they know someone to recommend, they won't look foolish when someone says, "Do you know someone I can call to open my car?" If they have heard your name and remember it they will refer you whether they know anything about your company or not! It is for this reason that referrals are so powerful. When someone recommends a company to us, we assume that company is good or our helper would not have referred them to us. This reduces the risk of "choosing the wrong door." The referral may even be made by a stranger. We still assume the company is reputable or the stranger would not suggest we use their services.

It is important to choose a name for your company that people can relate to. Personal names work very well such as: "YOUR first or last NAME" CAR OPENING SERVICE.

If your name is the same as someone they know, they assume they know you also (not actually, but the familiarity is there). It also feels more personal. Some people prefer to deal with an individual; others prefer to deal with a company.

It really does not matter whether you choose a personal name or a company name. However, if you choose a company name, please **do not** try to come up with something no one has ever heard of! You can use another company's name, if you like, if they are not incorporated, and if they are not in your town. Check out how many companies have the same name but are not owned by the same person. There are many because familiar names are reassuring.

The only problem with choosing a name that is already out there is that the first person to incorporate their business with that name owns that name for the entire state. As soon as one person owns a name they can make everyone else with the same name stop using it (if they want to). Choosing a name that people are already familiar with will put you way ahead of the competition. It doesn't have to be an exact copy of a name; something that sounds like something else will do.

If people hear your company name and think, "Oh yes, I've heard of your company," you've got a winner!

✍ Reminder

Step 1 Decide what you are going to call yourself

Step 2 See if a license or permit is required

Step 3 Choose a name for your business

Step 4 Register Your Trade Name/Tax Licenses

Now that you have chosen a name for your business the next thing to do is to make sure that no one else in your town is using that business name. We'll refer to the name you've chosen as your "Trade Name" which will also be known as your "DBA." This is short for "Doing Business As" and is the proper terminology for the name you've chosen to do business under. The way you find out if the name you've chosen is available for you to use, or if it is already being used by someone else, is to have a trade name search done. A trade name search is done at no charge and while you wait at the Department of Revenue in your town. If the trade name search comes up clear and no one else is using that name in your trade classification (you see, if someone else is using that name but not in your trade classification, you can still use it), then you can register your trade name. Examples of "same name, different classification" are:

■ Big Boy Lawn Service
■ Big Boy TV Repair
■ Big Boy Tire Center
■ Big Boy Records
■ Big Boy Etc., Etc., Etc.

Once you find a name that you can use, you will need to register your trade name with the Department of Revenue so that no one else can use it. You can register your trade name at the same office and normally with the same person who has done your search for you. It's just a matter of filling out a form and paying a small fee. Your Trade Name Registration confirmation letter from the Department of Revenue will arrive by mail in about two weeks. You will need to take this letter with you when you open your business checking account. You will have to open a business checking account in order to be able to deposit checks or money orders that are made out to your company name.

You are providing a service (considered labor), and while you are at the Department of Revenue you need to check to see if your city or state requires you to charge taxes on labor. If it does, you will also need to apply for a tax license. Actually you may need two of these, one for the city and one for the state. You can get the state license at the Department of Revenue while you are there. You'll have to go downtown to the city offices to get the city tax license.

We suggest that you go ahead and get both of these tax licenses anyway (city and state). If you decide you want to sell products at some time in the future (and by the way, you can sell products that have nothing to do with the business you are in), with these licenses you will be able to do so. Also, when you buy your products for resale, you will not have to pay taxes on them yourself.

Reminder

Step 1 Decide what you are going to call yourself

Step 2 See if a license or permit is required

Step 3 Choose a name for your business

Step 4 Register your trade name/tax licenses

 # Step 5 Business Cards, Envelopes, Stationery

The next thing you need to do is get some business cards printed. This will be **the most important tool in your possession**. You will need to become a professional business card dispensing machine!

STOP HERE! Before you get your business cards printed you need to read the next three sections:

1. **Read Section 2 on Advertising** for tips and tricks on how to design your business cards. Remember, this is your most important tool and we have suggestions that will make this tool very powerful!

2. **Read Section 3 on Communications** because you will need to choose a phone number to print on your cards. Knowing how you want to handle your incoming calls should be decided at this time.

3. **Read Section 4 on Insurance** to decide if you wish to be bonded and insured and if this is something you would like to advertise on your cards.

 ## Reminder

Step 1 Decide what you are going to call yourself

Step 2 See if a license or permit is required

Step 3 Choose a name for your business

Step 4 Register your trade name/tax licenses

Step 5 Business cards, envelopes, stationary

At the same time you are having your cards printed you should have envelopes printed with your business name and address in the return address area. You will also need a letterhead with your business name and address printed at the top of the page. A letterhead is simply a blank 8.5″ × 11″ sheet of paper with your business name and address printed on it. You will use it to write letters to suppliers and customers. Many companies will not open an account for you if you do not write your request for credit on your company letterhead.

Don't spend a lot of time trying to lay out designs for your cards, envelopes, and letterhead. Just go right down to a print shop and tell them what you want according to the guidelines we've given you. They have all kinds of ready-made artwork that you can use and will sit down with you and help you design these things. If you try to do it yourself you're going to knock yourself out, and before long you won't know if you like what you've come up with or not! The people at the print shop can show you how things will look on a computer before you agree to have it done. The whole process can take less than an hour!

CONGRATULATIONS!

If you have completed the first five steps, by the power vested in your state with the issuance of tax licenses and trade name registration (and operating licenses or permits if required), **you are now officially a locksmith.** The documents you now possess are considered proof thereof.

NOTE: Your cards, envelopes, and letterhead are sometimes all the proof required by suppliers and are normally only necessary with your first order. It is not necessary, nor is it recommended, that you tell anyone that you have no experience at all. A good answer when someone asks you how long you have been doing this is, "Oh, for a while now."

Now that you have become a locksmith you can order your tools and begin your training. In the section on tools we have explained the differences between three of the best tool companies on the market and the training methods used by these three tool manufacturers. You should have already read the sections on **Advertising**, **Communications**, and **Insurance**. Don't confuse advertising with getting business; you should not have read the section on **Getting Business** yet. Now let's read the section on Tools and order our tools!

Step 6　Order Tools, Books, Videos

At this point you should have ordered your tools. Thoroughly examine and study your tools and any books, tapes, or videos that you have ordered. After you have studied these you will need some hands-on experience and practice.

For information on how to get this: **Read the section on Training and Education**.

Step 7　Develop Your Skills and Practice

There are a few more things we need to do before we get our first paying customer! The next two things we need to do are to learn about the paperwork that we will need to run this business and how to charge and collect our fees. The section on **Paperwork** will show which kinds of invoices to use and other forms associated with this business. The section on **Money Matters** discusses how to charge for your services and the different methods of payment you may wish to accept.

Step 8　Develop a Customer Base

Let's make sure we look the part by reading the section on **Appearances**. If you have your invoices printed you are ready for customers. Wait— remember, if you are going to have checks made out to anything other than your personal name you will have to have a business checking account open in your business name to cash those checks.

We need to be prepared for any problems that we might encounter so let's read the section on **Problems** and then go right into the section on **Getting Business**. Now we can start handing out those business cards and building our business!

Step 9 Have Fun and Make Money

This has just been too easy. The next section to read will be **Telephone Skills** and this is very important. Practice your telephone skills and always ask yourself after every phone call that ended without a sale, "How could I have spoken to that customer differently that would have made them decide to use my service?" This will take some time. Have your friends call you pretending to be customers. This will help you feel more self-assured when you begin talking with real customers.

The remaining sections of the manual can be read at your leisure. Your business should be up and running by this time. The section on **The Law** is general information. If you received a rules and regulation package from your city or state, it will have more specific laws that relate to your area.

The sections on **Trade Associations** and **Schools** are for continuing your education. We recommend that you join one of the associations we have listed, or join a local group in your town. There is a chart in the **Schools** section that shows you which schools are home study or correspondence courses. If you think you might want to take one of these courses you need to contact the school now and ask them to send you information. By contacting them now, you will begin getting coupons that will eventually add up to $350 off one of the courses.

It's A Fact!

The more you learn, the more you'll earn!

We have also included a list of **Suppliers** to give you many sources to order your tools and supplies from. Sometimes just finding these contacts is the hardest part of starting a business.

We have given you enough information in this manual not only to get you established in the car opening business, but also to launch you into becoming a full-fledged locksmith should you choose to do so.

Step 10 Talk to a Bookkeeper or an Accountant

As soon as you start earning and collecting money, you will need to seek out professional help to keep you from getting into trouble with the Internal Revenue Service. A competent bookkeeper or accountant should be found.

These professionals should know the rules of business and can answer any paperwork-related questions you may have about your business.

This business is very simple, the rules are also simple, but you need to know what the rules are for your city and state. The IRS says, "Ignorance is no excuse."

 WARNING: The IRS will not let you off the hook just because you were not aware of the rules!

Drop us a note when you are successful. These are the best letters we get. Work smart, make lots of money, and we continue to wish you good luck!

Advertising

SECTION OUTLINE

SECTION 2

Advertising

Introduction

The purpose of advertising boils down to just one thing, getting more business and opening as many cars as you can at a profit. The better your ad is, the more profit you will make. If your advertising is correctly targeted to the car opening market (in the places where people will look for help when they get locked out of their cars), the more profit you will make. Another factor to consider is the number of prospective customers you will reach and how your advertising is presented to them. Of all the things that you will do in your business, advertising will play the most important part in getting new business and referrals.

You've seen or heard catchy ads on the radio or TV and seen good and bad business cards. You probably even have some type of advertising in your possession right now. Stop and think; look in your wallet or purse; see what you have. Now ask yourself why you still have it. Maybe you need the service or you're saving a discount coupon. Keep in mind that this is what advertising is all about: Getting your prospective customer to remember you and not your competitor.

 ## Advertise Professionally

In the car opening business you will not meet anyone who has not needed your services in the past or who might need you to open his car door, trunk, or gas cap sometime in the future. Just mentioning that you are a car opening expert is a form of advertising.

There are many ways to advertise and just as many ways to pay someone else to do this for you. Your own effort will pay for itself many times over. Word of mouth is by far the best advertising that you will ever have. But will your prospective customers remember your company name or phone number when they are in a frenzy to get the keys out of the car? You need to have different types of advertising.

It is important in this business that you do not use cute, clever, or funny advertising. This will be a handicap to any advertising that you do. You need to remember that this is a unique business. People develop emotional attachments to their automobiles and most people want a professional to open their car for them. Silly advertising does not look "professional" and it does not reassure your customers that you are a professional.

The following points should be part of all your advertising. Your ad should:

- Attract attention.
- Develop interest.
- Describe the service you are offering.
- Convince the customers that they should use your service.
- Tell how to order or receive your service.

For an effective ad you need to make sure each word counts. Avoid unnecessary words, and put as much action in your words as you can. Use simple terms that your readers will understand: the simpler the better. Get right to the point. Don't beat around the bush. Only use flowery statements when they don't get in the way of the point you are trying to make. Keep in mind the reason that you are advertising is to get people to call you to open their cars when they get locked out.

Be very consistent with all of your advertising. If you have a picture of someone opening a car in one of your ads, then put that same picture in all of your ads. This is called product, or (in our case) service recognition. Some examples of ads which will help your service recognition are business cards, yellow pages, brochures, and flyers just to name a few. Be sure to be consistent—these ads are seen by almost everyone sooner or later.

Simple layouts are the best and be sure that your advertising flows well. Lead your reader's eye easily through the advertising from the headline to the art work, to your message, then to the price, and finally to the phone number.

Keep in mind that your customers only have one thing in mind, "What's in it for me?" Give your potential customer one primary reason why he should use your service, then tell him about all the benefits. Benefits are the reason why people will choose your service. A benefit could be that you can respond faster, or that you are the very best in the business, or that you offer a discount, and so on.

 ## Spending the Buck

Many small business people start out with ideas and attitudes that are penny-wise but pound-foolish. Knowing how to cut corners can be very helpful, but only up to a point. Don't be afraid to spend money on basic investments.

The trick is to know which expenditures will pay off. To give you some idea of this, think about the high-end purchases that a car opening professional really does not need to make to start a business. Computer systems, new service vans, or a shop would be extreme overkill. What you will need are service manuals, tools, and advertising. The best way to determine what might be important to your business is to ask yourself:

Will this save me time?

Will it make my job quicker?

Will I make more money by doing this?

 # Business Cards

Your business cards will sometimes be the first impression that your new customer will receive about your company. The best cards that we have seen are the ones that convey a little extra information, like your slogan or your company message.

In designing your business cards there are several things to keep in mind. Limited space, vital information, visual recognition, and a reason for keeping the card are important. Since your business card is so small, the impact of your card is very important. Impact can be many things: color, size of print, or the layout of the graphics. The main idea is recognition. Remind your customer who helped them get back into their car.

Vital information: Your company name, address, city, state, zip and most important, your phone number. With all this in mind you must keep the card attractive and affordable. It is important that your business card have the following information on it. If you are mobile only, your address is not necessary.

- ☐ 1. Your business name and address
- ☐ 2. Your phone number
- ☐ 3. What it is that you do
- ☐ 4. Your business hours
- ☐ 5. Your message or your normal response time (15–20 minutes is good)
- ☐ 6. A reason why your customers should keep your card
- ☐ 7. If you are bonded and insured.

If you can get all of this information on your card, it will be a very powerful tool.

Should you want to run your business on a part-time basis, simply state on your card what your hours are. For example: 6 p.m. through 6 a.m. Mon.–Fri., 24 hrs. Sat. & Sun. If you are going to do this full time, it is best to say 24 hr. mobile service. If you are willing to make your card a discount coupon for, let's say, 10% off each time your card is presented, people will be much more inclined to keep it. This is also a good source for referrals. Someone holding your card may have the opportunity to help a friend or

stranger by giving them your discount card. It gives them a chance to be helpful and it makes people feel good.

Visual recognition: When your customer sees your card they instantly remember you or your business. This can be from seeing a picture of what you do or how fast your service is, but whatever it is keep it simple, the simpler the better.

Colored cards: When printed effectively, colored cards are very striking and will give your company the appearance of being successful.

Who should you give your business cards to? Everyone you meet, plus many more. Each time you hand out your card, hand out three cards. Ask your prospective customer to give two of them away. You will be surprised at the number of people who will do this for you. If you give several business cards away several times a day, within a short period of time you will have so many cars to open that there won't be enough time to do them all. The key to using business cards is getting your customers to keep them.

Of all the information on your card, the phone number should stand out the most. If people can't read it or the print is too small, they probably will not keep your card. Keeping this in mind, you still need to give them a reason to keep your card. One good way is to give them something for free. This could be a small gift such as a key chain or a discount for using your service. The company that I own gives an increased discount value each time the card is presented until the customer has reached 50%, and then we start again at 10% with a new card. This is printed on the back of the card. We date and sign the card every time it is used (every time a customer locks the keys in the car). The reason to keep the card is the discount coupon printed on the back of the card.

 # Magnetic Cards

Magnetic cards are very valuable when used correctly. You don't want to hand these out to everyone you meet because they are expensive. You will give these to store clerks and used car dealers, hotel clerks, and anyone who will display your card for quick reference. Store clerks will place this on the cash registers to help customers, and used car dealers will want this close by for emergencies. Give this type of card to people that you believe will use it; it won't do you much good if it is used as a refrigerator magnet.

Having magnetic cards made is expensive and you don't need to do this. At the print shop where you are having your cards made, ask for magnet strips that stick to the back of your cards. This is a new product that makes magnetic cards inexpensive and they look the same as the real thing. Start with around twenty-five of these; then if you need more, return to the print shop for more magnets.

 # Rolodex Cards

Rolodex cards are cards that fit into a card file. This is your basic business card only larger and it is an excellent item to stuff into your statements when you are mailing them. Give these to customers that you are meeting for the first time and your old customers that you have had for a long time. When you print these cards with all of your important information, your customer does not have to do a thing except stick it in a card file. This is a genuine benefit to them, which makes it a genuine benefit for you.

 # Yellow Pages Advertising

Prerequisite

Before you can have an ad in the yellow pages you will have to have a business telephone line. By this we mean you must either contact the phone company to have your existing home telephone number upgraded to business class or, if you already have a second line, you must be paying business rates on that line. There is no difference between these two phone lines except for the fact that you will pay about twice as much for the business line. You cannot get yellow pages advertising unless you are paying a business rate on one of your phone lines.

Why the Yellow Pages Advertising?

Sales, sales, and more sales. This type of advertising is very important because this is the main source that people use to find help after everything else fails—After they have tried to call someone they know for help or get someone close by to help them. After they decide that they cannot open the car themselves, then they will look in the yellow pages to find you.

The first thing that some people do is pick up the phone book and look in the yellow pages for help. Not everyone will go to the yellow pages first, but many people do. Probably the only people who won't eventually use the yellow pages are the ones who already have your business cards and people who are referred to you by others who are trying to help. Remember, car dealers, store owners, and your customers are eager to refer others to you; everyone else needs the yellow pages.

What Needs to Be in Your Ad?

The same basic things that are on your business card plus a few more things. You could elaborate on your services, tell your customers how fast your service is, and give your hours of business. Tell them if you do emergency work, if you have 24-hour service, what areas you cover, and maybe even tell what types of payment that you will accept.

What Size Ad Should You Get?

The size of your yellow pages ad will make a difference in the number of calls that you get. A full page ad will get many, many more calls than a single one line ad that just gives your name and phone number. You will need to check the prices on the different sizes and choose a price range that you don't feel will keep you from sleeping at night. Yellow pages advertising is expensive. We strongly suggest that at the very least you choose an ad size no smaller than a dollar bill. We call this a dollar bill size ad because it is the same size as a dollar bill.

A quarter page ad will be even more effective but you may want to start a little smaller the first year until you get comfortable with your new business. You can increase your ad size each year until you end up getting the number of phone calls that you want. Your yellow pages representative will sell you any size ad that you want but keep in mind that they do business on a first come/first served basis depending on the size of the ad. The larger ads are placed in the front of the smaller ads. This means that if you buy the biggest ad, you are going to have first position under that listing, which should be in the Locks/Locksmiths section.

Location Makes a Difference   Most people who realize that a locksmith is someone who opens cars will look in that section of the phone book but there are also other areas in the yellow pages where your customers will look. Some examples of these areas are: keys, towing companies, emergency road service listings, car dealerships, and gas stations. You should start with your advertising under the heading Locks/Locksmiths and then, if you wish, experiment in the other areas with smaller ads to test their effectiveness. We are currently under the heading of Locks/Locksmiths with a half page ad and with only a one line listing under the heading of Keys. We've found these two listings to be very effective and are all that we need. You must realize that if you advertise your lockout service business under the towing section you will be spending a lot of your time explaining to people that you do not tow cars.

Oops! You Made a Mistake   Sometimes what you thought would work does not. Example: you have a large picture of a buffalo in your ad because your company name is "Buffalo Car Opening Service," when you should have had a picture of a car in there instead. Therefore, you may not get great results even with a very expensive ad. If you are going to go for it and get a very large yellow pages ad right from the start, the rules in our town are: If you find you cannot pay for your yellow pages advertising because it is not pulling the results that you expected, you are not liable to pay for the remainder of your advertising if you have the phone number in the ad disconnected. This goes into effect from the time you disconnect that phone number and is not considered bad credit. The balance of the advertising is not used and you don't pay for what you don't use.

Your representative will not tell you this up front; you will have to ask if these rules are the same in your town. We are only telling you this to make you feel more comfortable about committing to such a large contract. If you have not advertised in the yellow pages before, the cost of advertising may surprise you. Keep in mind, however, that the larger your ad is the more calls you will get.

Check your current phone book to see how the other locksmiths are advertising. You don't need a half page ad if all the other locksmiths are only dollar bill sized or smaller. Adjust to your competition. If all of their ads are small you would be throwing money away buying a half page ad. All you want to do is to get close to the front of the listings. The closer the better.

How Do You Get Started?

Find your local phone book and look in the first few pages of the book for directory advertising information or call the operator and ask for directory advertising. The operator will give you the correct phone number.

When Should You Start?

As soon as possible. The phone books only come out once a year so, if you've missed the current deadline for the coming year, you will have to wait until the following year. The phone book has a deadline on when you need to get your ad in. Each city has a different deadline so be sure to contact a sales representative right away. Keep in mind that you will be signing a contract for one year. Once you have an ad, a representative will contact you long before the next year's deadline so you will have time to revise and/or upsize your ad.

What Kinds of Help Can You Expect?

Some representatives will only do what you ask of them. If you can tell that you have a representative who doesn't seem interested, ask to have another representative assigned to you. When you get a good representative, and if you have time to play with before the deadline closes, they will do everything for you, and we mean everything. Just give them your company name and phone number, tell them what you do, forms of payment you will accept, and the hours you are open; and ask them to design an ad for you. They are trained for this and know what a good ad should look like. They will do all the art work for you and lay out an effective ad. It normally takes about four weeks to do this and (if there is time before the deadline) the representative will do this several times until they find something that you like. They will show you what is called a "proof," which is exactly what the ad will look like in the phone book. You are not legally bound to your contract until you approve of and sign this "proof."

If you are short on time before the deadline, you will have to be more specific about what it is that you want. The representative will provide testimonials from other advertisers in your classification and will show you many other ads from different places to see if you like any of that art work or would like to have something similar placed in your ad. Remember that you are not being charged for all of this design and layout work. This is a free service that the company provides, so use their services as much as possible. You only pay for the size of your ad.

Small Phone Books and White Pages

Here are some suggestions if there is more than one phone book in your town. Small phone book advertising is one area not many people in the car opening industry use or take very seriously. This advertising is cheaper and produces better leads than the average car opening expert would expect. If you get into one of these books and are the only locksmith advertising in that book, who do you think is going to get **all** the calls? The reason why these ads are less expensive is that they do not cover as large an area and have a smaller audience than the larger phone books have. Some of the little known areas that should be looked into are your surrounding small town yellow pages. Local air force and army bases, larger home owner associations, and smaller newspapers are just a few of the different types of niche advertising markets available to you.

White pages are good if your customers already know who you are and what you do, but don't help much if they are looking for the first time. Bold listings in the white pages will help your customers locate you in the white pages but only if they are looking specifically for you.

Phone Book Covers

Just after the new phone books come out you will get many calls from people who say they print phone book covers and are looking for only one advertiser in each trade classification to be printed on their covers. This sounds tempting. You would be the only locksmith on this phone book cover and that sounds like a good deal. This is not effective in our business. The quantity of distribution of these covers is too small to be cost-effective advertising. When advertising, always look to the places that people will look first.

We don't do this type of advertising for the simple reason of economics. We try to keep the advertising dollar in the areas of the phone book where most people will look first. These areas are the primary heading of Locks/ Locksmiths and the secondary heading of Keys. From these areas we get the majority of the lockout calls that come into our businesses.

Specialty Numbers

Ask your yellow pages representative for information on getting an 800 number. This is a fairly new service being provided at a very low cost. You only want to advertise this number in your yellow pages advertising and nowhere else. Not on your cards or on any other advertising that you are doing. The purpose of the 800 number is so that when someone is locked out of their car and shopping the yellow pages, they can use the pay phone to call you for free! (You will be charged for this call.) In fact, this should be one of the selling points in your yellow pages ad. Example: "If you're calling from a pay phone, let us pay … call 1-800-555-1234." This would be a separate number in addition to your regular number.

Some car opening companies use pretty catchy phone numbers that are easy to remember such as: 555-OPEN or 555-LOCK. If you can come up with one of these it would be beneficial and would help people remember your number, but this is not a critical factor to your success.

 Flyers and Direct Mail

This form of advertising is inexpensive and can be very profitable if you direct it to a target market or groups of customers that use car opening services. This form of advertisement can be tailored to fit a group or area and the type of customer that you want. Be sure that you get professional help on the art work and keep things as simple as you can.

Keep in mind these basic rules and ideas when doing your direct mail flyers. Send to a predetermined and targeted group of clients. Be sure these people meet the criteria of your objective. Sending to a new car dealership might be a waste of your time and money. On the other hand, used car dealer contacts can be very profitable. Customize your mailer to fit your customers' needs. Don't try to sell them tires if you really want their car opening business. Speak directly to them and fill a need. Include some type of special offer: a coupon, a free gift, or the first opening free. This will give them a reason to call you back. Design with the purpose of getting them to remember your name. This could be done with a slogan, some eye-catching logos and graphics, or printing your phone number big and bold along with your company name. Follow up is important if you want to get the account.

The basic reasons you should use flyers are that the cost per each piece you send out is relatively small and that you can target a very narrow clientele. By this we mean you, as a car opening expert, can pick the type of businesses that you think will be most beneficial to you. Examples are car rental agencies, used car lots, and even emergency road service companies.

The flow of your flyer should be from some type of graphic that represents what you do to text that tells your story in very few words. This will tell your potential customer what your company does without them working very hard to figure it out.

The graphics should be simple but very clear. Be sure the graphics make the point that you're giving them a professional looking flyer. Good graphics will also make clear what you are trying to sell them. Color is not as important but does help draw more attention to specific areas in the flyer. Be very specific about the topics that you choose. Draw them to a special in your flyer if at all possible; this will give them a sense of urgency to call you.

You will need another flyer made that is tailored to the general public to help build name recognition. This type of flyer can be placed on car windshields in parking lots or handed out door to door.

Newspapers and Free Advertising

Newspapers are another form of name recognition advertising but will not cause immediate measurable increases in your business. News releases are a form of free advertising. The hard part is finding something new and interesting that the newspapers want to print. Just because you think that you have the newest and best widget does not mean that the editor of the newspaper will think it is great. Smaller newspapers will be your best bet when it comes to news releases as long as your story is not presented as self-benefiting. A story about the costs of breaking a window versus paying someone to open your car could be a free news release story.

By giving away your services to charities, you not only help your community but also create a positive awareness of what a car opening expert does. That type of advertising can't be purchased anywhere for any price.

Signs on Your Vehicle

This area of advertising is a must. You should have signs on your vehicle as well as on your place of business if you have a shop. Don't be cheap when it comes to your business image. A ratty looking sign will reflect poorly on your business. Of all the advertising that you do, this is one place where you should spend the extra money to get a professional job done. Let the quality of your sign reflect the quality of work that you do. Be original and use your logo, and be sure to tie all this in with the theme of your business.

Radio and Television

Television is not recommended when starting out because of the expense. TV gets about the same results as radio: good for name recognition, not for immediate sales. Start in areas where you can measure immediate increases in your business and then work on name recognition.

Radio follows listeners everywhere in the home and on the highway. It's known for lower rates compared with other types of advertising. In the field of car openings, it may be good for recognition but won't do a lot for emergency lockout services. If you are trying to have your customers recognize your name when they need you in the future, this is a good way to start.

 # Referrals

Referrals are one of the best forms of advertising that you will never pay for. This form of advertising is done by your good will and from quality work that you perform. If you can get one referral from each customer you help, you will never need any other form of advertising.

Communications

SECTION OUTLINE

SECTION 3

Communications

Introduction

Communications is a very important area that needs to be addressed. There are many ways that you can set up your telephone answering system. If you do not do it properly you could be cutting your business in half. We see this being done wrong every day and all we can say is—more for us!

Your telephone answering system is simply the way in which you handle your incoming calls. In this section we will tell you exactly how to handle your incoming calls for optimum performance (getting as many jobs as possible).

There are many communication devices available to choose from such as commercial telephones, residential telephones, cellular phones, radios, radios that have telephone capabilities, pagers, answering machines, answering services, answering services that will talk to your customer and then page you, and even computers that have answering machines within them. There are also many features available to go with these devices such as voice mail, voice mail that will page you when you have a message, call waiting, call forwarding, caller ID, last call return, and forward if no answer.

Having all of these options available can make things very confusing. If you will keep in mind that you are operating an emergency service business and put yourself in your customer's position, things will become clearer. When your customer is locked out of their car at a public place, they are embarrassed, and will remain embarrassed until the problem is taken care of. This is where most of the *urgency* comes from. This is a very uncomfortable situation and the customer wants it taken care of as quickly as possible. For this reason, 90% of your prospective clients will not leave a message on an answering machine or wait for someone to call back. The client also may be calling from a pay phone and unable to leave a number where they can be reached.

 Cellular Phones

Clearly, the best way to handle incoming calls is to personally answer each and every call yourself. There are two methods that work well for this. One is to use radios which we'll cover a bit later. The best way to set up your answering system is to advertise a phone number for your business that is a phone line going into your home. You can use your home telephone number or you can have another phone line installed. We recommend having a second line installed in order to keep your personal calls and business calls separate. This will ensure that the business line is always answered in a

professional manner. If you have another phone line installed, you will need to purchase a two-line telephone to accommodate both line one and line two. The phone line that you are going to use for your business needs to have call waiting and call forwarding activated.

As your business grows you will begin to notice that your lockout calls are coming in groups of two or three calls at a time and you will need call waiting to capture them all. Call forwarding will allow you to leave your home to open a car without shutting your business down until you return. You will need to forward your calls to a cellular phone when you are not at your home if you want to continue to receive calls. Choose a cellular phone that clips onto your belt instead of a bulky one that you carry in a case.

There are two options that you will need to activate on your cellular phone. They are call waiting and voice mail messaging. Call waiting works the same as it does on your home phone. If you are speaking to someone and another call comes in, you will hear a beep notifying you of an incoming call. Ask the person you're speaking with to hold and you can switch between the two parties.

When you get a call on your cellular phone, if you do not answer within a certain number of rings, it can automatically transfer your caller to voice mail. There will be times when you are in an area where your cellular tower signal cannot reach your cell phone. At these times your calls will go directly to your voice mail. If you have the option of having your cell phone notify you when you have messages, it is worth having this feature activated. Most cell phones do have this option.

This is the standard voice mail message that we use for our customers. We've found this message to be better than anything else we've tried in order to get someone to leave a message … "Hello, you have reached 'your business name.' We are being alerted to your call. Please leave a message and we will return your call as soon as possible. Thank you."

"We are being alerted to your call" gives the impression that we will call them back immediately, and we have found this phrase to be the most successful in getting a potential customer to leave a message.

Voice mail messaging is also convenient for those times when you just do not feel like answering the phone. It is always better to have a customer reach your voice mail than it is to just let the phone ring. Occasionally, if someone is calling around for the best price, they will leave you a message.

This is the way we run our businesses. We have a residential phone line, and in addition we have a business phone line. The residential phone line is for mom, the kids, and calling the relatives. The business line is connected to the same phone (a two-line phone) and the business line is strictly for business use only. We have call waiting and call forwarding on the business line which lets us forward our calls from the business line directly to our cellular phone. We also have voice mail messaging on the cellular phone so that if, for some reason, we don't answer the cellular phone within four rings, the call will go directly to an answering machine (voice mail). To give you

⌕ Reminder

A business line is for business use only.

an idea of how we work our phones: we get our first phone call on the business line at home; after that, calls are forwarded to our cellular phone. This system ensures that we won't lose any calls. We use our cellular phone every single day, day in and day out. It's probably one of the best tools we have. And that's what you have to consider it—a tool.

One of the drawbacks of a cellular phone is the cost. Cost is a huge factor with a cellular phone. You not only have a monthly service fee, you also pay by the minute. An example of the cost involved in using a cellular phone is as follows. In this area we pay $30 per month just to have the cellular phone number. Then we pay 48¢ per minute for every minute that we use the cellular phone. Most cellular services charge by the whole minute. If you talk to someone for two seconds on your cellular phone, you are charged for the full minute. We make enough money off each service call we make as a result of a cellular call to more than pay for the cellular phone charges.

You may be tempted to advertise your cellular phone number as your business number to make things less complicated. Do not give in to temptation. If you do, your cellular phone bill will be very expensive and most people can recognize a cellular phone number prefix. When customers see a cellular prefix they will have the impression that you are working out of your home. This will damage your professional image. It will be more expensive for you and you will get fewer calls if you appear to be a "home-based" business. Your image is very important to your business.

Radios

The next best method is to have someone at home answer your calls and dispatch you by radio. When I first started out, I used two-way radios and my wife answered the phone for me. Back then, two-way radios worked out pretty well if my wife was at home. The cost was really inexpensive; we purchased used radios so we didn't pay very much for them at all. Here's how our system worked:

The telephone call was answered at home.

My wife would then call me on the radio and say, "When can you open a car," or "What part of town are you in," or "I have a car for you to open downtown."

I could pretty much respond instantly.

She would give my response time to the customer right away.

This was good for starting out.

Probably the biggest plus for radios is the cost. Air-time charges for radios are normally a flat fee per month for each radio. We only spent $35 per month to use both radios and we used the radios all month long. There is a drawback to using radios. When your wife, or whoever is answering the phone for you, is speaking to a customer on the telephone, the customer can also hear you on the radio through the telephone. This sounds kind of odd, but there were many times when I would have liked to say, "No, I don't want that job because …", but I couldn't say it over the radio because I knew the customer could hear me.

Answering Services

Ace answering service

Answering services have a place in your business and are mainly used when you need some time off. They have some good points and some bad points. Of course, the obvious good point is that you actually have a physical person answering the phone. Unfortunately, this real person doesn't know what your prices are or how soon you can respond. You will lose business if you are answering all of your calls through an answering service. People do not want to wait to find out **if** you can come open a car.

One of the good things about an answering service is that, at night when you get off work, if you want to stop at 6:00 p.m. so that you can go do some of those extracurricular activities, you can call forward your phones to the answering service. The service will take messages so you won't physically have to answer the phone. It is always good to have a person answering the phone, but you do not have to accept every job that comes in.

Answering services are a little bit expensive. You may not want to use an answering service right away, or you may not want to use one at all; but there will be times when you will probably want to have one, especially if you happen to go on vacation. You don't want to just let your phones ring off the wall, unanswered, for a week straight. If you do let your phones go unanswered, your customers will think you're out of business. At least with the answering service, your customers will be told that you're out of town, or that you will not be back in until Monday morning. You don't have to hire an answering service for a full year. You can use them specifically for when you go on vacation, or just at night, or for emergencies.

Pagers

Pagers are a form of communication that sounds good but does not work well in this business. There are quite a few drawbacks to pagers. Probably the biggest drawback would be your response time. When customers call for emergency service to unlock a vehicle, they want to speak with a real person. They don't want to dial in a phone number and wait five or ten minutes to have somebody call them back. They want to talk to somebody right now.

Now, pagers are cheap. You can probably buy and own a pager for $30. We've seen specials on them and air time is probably about $5 per month. However, even though it's cheap, you're going to lose business because you cannot speak directly with the customer. You could have your calls go into an answering service and then have the answering service page you. You will get the message real fast, but you will still have to stop someplace, find a pay phone, stick your quarter in the telephone, and call the customer back. Well, that's too long for a client to wait. Your customer doesn't want to wait more than just a few seconds to find out if somebody is going to be able to help.

Pagers may sound like a great idea and they may be cheap, but be sure you check out those drawbacks. We don't recommend pagers at all, especially if you want to try to make a living in the car opening business. **You will probably lose 80–90% of your potential customers just because they do not want to wait for somebody to return their phone call.**

Caller ID

You are probably already familiar with Caller ID. If you want to have this service so you will know when your competition is price checking you, it is probably useless. Anyone who does not want you to know who is calling can block their number by dialing two extra numbers before they call you. We were excited when this service first came out but we canceled it after it proved to be of no particular benefit to us. You may want this feature so you will know when you are getting a late night call from a nightclub or bar, but other than that, **save your money.**

Insurance

SECTION OUTLINE

Insurance

Introduction

In this section we will give you a basic idea of the different kinds of insurance that are associated with the car opening business. The insurance that we are covering in this section is **extra** insurance that you may want to have to go with the insurance that you already have or may be required to have.

We don't claim to be insurance agents. We are not insurance professionals so, whatever you do, you always need to consider talking to an insurance professional about your business needs. Every state has different requirements, and insurance is very important. Some states require that you carry certain types of insurance. Other states don't require any form of insurance. You should carry the insurance that you need to protect yourself, your van, your business, and your family.

The only reason that we have a section on insurance is because most locksmiths advertise that they are **Bonded** and **Insured** and you are probably wondering what that means.

 Bonded

What does bonded mean? Most locksmiths advertise that they are **bonded** and many customers look for this in your advertising. Most people think that in order to get a bond you have had to go through an extensive background check on your character and that only the elite in both honesty and trustworthiness can get a bond. This is good for your business.

A bond is simply insurance for your customer. It guarantees that money is available to reimburse your customers should you be convicted of stealing from them. In this business you are not required to be bonded. It is simply an option that may benefit your business by bringing you more customers. We recommend that you become bonded.

It is very easy to get a bond in this trade. Some trade magazines offer bonding at a very low price. Fill out the order form that is in the trade magazine that you've purchased, send in the fee, and in about three weeks your bond certificate will arrive in the mail. There are no background checks, and nothing is required other than being a subscriber to the magazine.

Each magazine has its own bond and they vary from a low of $5,000 to the highest (that we are aware of at this time) which is $15,000. The amount of the bond is the amount of money that is guaranteed to be available for reimbursement to your customers if you are convicted of stealing from them. The cost is very low, $5 to $15 per year. The bond normally expires with your magazine subscription. For more information on bonds and magazines see the section on Training and Education.

Insured

What do they mean, exactly, by insured? Insured means that you are covered by commercial general liability insurance. This type of insurance covers things like property damage and personal injury. If you are working on a job and you make a mistake that causes someone to get hurt, or if you damage property, you will pay your deductible and the insurance company will take it from there.

Now, some states have different regulations and some insurance companies offer different kinds of coverage. Be sure you talk to your insurance agent about exactly what your policy covers. Ours will cover damage to personal property, "false" advertising (if we accidentally advertise something incorrectly), fire damage, any kind of medical expenses, products, installations that were incorrect, and so on. Understand that "false" advertising and incorrect installations are only covered so long as we intended to be fair and honest, and only if we were not incompetent in our installations.

Commercial General Liability Insurance is not very expensive for the coverage that you get as compared to other types of insurance. In our area it runs about $400 per year and is paid in one lump sum payment. It is well worth the price.

Workers' Compensation Insurance

This insurance is mandatory if you have employees. The only time that you do not have to carry workers' compensation is if you **work by yourself** and have **zero employees**. Subcontractors are considered employees in most states and have to either be covered or carry their own coverage.

Tools

SECTION OUTLINE

SECTION 5

Introduction

At the end of this section we have made a list of all the tools, books, and videos that we believe you should order to get started in this business. You can save money by choosing cheaper products or economy sets but be careful. Trying to save money on tools could end up costing you money on a daily basis. If you cannot afford to purchase the tools that we are recommending, you should make that purchase one of your future goals.

You should always have a backup system in place—whenever possible—in everything that you do in this business. This is one thing that makes the difference between a professional and an amateur. You will want to have more than one method of opening each vehicle available to you whenever possible. This is a deciding factor when we are choosing and recommending car opening tool sets. One of your car opening backup systems will be your lock picking skills. Most automotive locks can be picked open quite easily.

It's A Fact!

A backup system makes the difference between a professional and an amateur.

Lock Picking Tools

Let's begin by looking into lock picking tools because this is where we believe the most confusion is for beginners. Looking through catalogs from your suppliers you'll find individual lock picks and lock picking sets for sale. The confusion comes from all the different shapes and sizes. There are some very fancy lock picking sets for sale out there. If you are a beginner this can be quite a challenge. If you have no lock picking experience you may be tempted to purchase a large set of lock picks just to make sure you will have the right tool for each job. You could easily spend $250 or more on one of these fancy lock picking sets when, in reality, you only need to spend $20 to get all the lock picks that you will need.

Lock picks come in all shapes and sizes but there are only four different shapes that you will need. They are:

1. Diamond
2. Rake
3. Hook
4. Double ball

Forget about all the other shapes for now. After you get some experience with lock picking you will know what all the other shapes are about just by looking at them. Basically all the other shapes are for customizing. You will develop your own personal preferences as you gain experience.

We use the diamond-shaped pick 90% of the time. If this were the only pick that we carried we do not feel that it would hinder our performance. The other three shapes that we use are primarily for our convenience. The double ball pick (a double-sided tool) is used only on double-sided wafer locks. Even then it is used only so you don't have to remove it from the lock to turn it over and then reinsert it back into the lock to pick the other side. No lock picking instructions will be given here because we don't know if you have decided to become a locksmith or not.

You will also need two or three different tension wrenches. We prefer a short, stiff tension wrench and we carry two others for odd-shaped keyholes. You will need to experiment to find out which tension wrenches you like best.

If you purchase your lock picks separately, the average cost of both picks and tension wrenches should be about $2 each. You don't need a 164-piece set. Just start with the four basic shapes—diamond, rake, hook, and double ball—and a few tension wrenches.

The final lock picking tool that you really must have is a pick gun. You will not use a pick gun on automotive locks because it can cause the spring cap cover to pop off. When that happens you will spend a couple of hours putting the lock back together. If you are only interested in opening car doors you won't need this tool. If you want to be able to open every lock that can be picked open, this tool is a must.

Which lock picking book should you order? Although all the books available are slightly different, they cover the same material. Lock picking is the same in one book as it is in another. There is no magic trick for picking locks; it all comes from practice. The only thing you will get out of these books is theory. You do need to understand how the different types of locks work to be able to pick them open and you will learn to recognize these differences with just a little practice. The only thing that will make you a lock picking expert is practice, practice, and more practice. We suggest you purchase one basic lock picking book and one lock picking video. Start with these two and then start practicing. If you feel that maybe there is something you missed in theory, order another book. A different book will be written in a different style and that may cause the light to come on for you.

 WARNING: Do not start this business thinking that you can save money on tools by relying solely on your lock picking skills to open cars. Some automobiles cannot be picked open. You must purchase a car opening tool set to run this business.

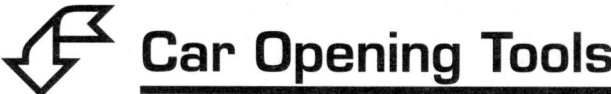 **Car Opening Tools**

There are many car opening tools out there and there are many car opening kits to choose from. Knowing which ones to buy can make or break your business. We've seen many car opening tool sets and manuals and, to be quite honest, we don't see how some companies continue to sell their products or stay in business.

There are several factors to consider when choosing a car opening tool set. These factors are:

Do they update their manuals and tools every year for the new cars?

How much do the updates cost?

Will this tool set open virtually every car on the road?

Do they only show you one way to open each vehicle or are there alternate methods available?

Are the manuals written in such a way that you can understand them and get the job done quickly?

We are going to take a look at three different car opening tool companies so you can see the differences. In our opinion these three companies are the very best in the car opening business.

Keep in mind that each of these companies offers economy car opening sets. Since these economy sets do not open virtually any car on the road, we don't feel that we need to discuss them here. You need a set of car opening tools that you can hit the road with when you get a job and not have to worry about whether or not you have the tool you need to get the job done. Any one of these three sets will work for you.

There are times when you need to shop for the best price. **This is not one of those times.** All three of these tool sets are in (approximately) the $200 price range. You are not looking to save money here by purchasing the cheapest set. You are looking for the best set you can get. Each company has a different number of tools in its kits; some will have ten different tools and some will have forty different tools. Do not let this affect your decision to buy. It means very little that a kit has forty tools if you will only end up using ten of them.

Tech-Train 2050 Master Locksmith Car Opening Kit

Our personal favorite tool set is from Tech-Train Productions. This is the set that we are going to recommend to you as first choice. The name of the set we use is *Tech-Train 2050 Master Locksmith Car Opening Kit.*

We chose this set because, with these tools, we feel that we can train a new employee to open cars much faster than with any other system available. There are nine car opening videos that are available with this system.

They cover every car in the opening manual. Each car in the manual is referenced to a video tape in the index of the manual so that, if you have trouble on a job, you can look up that vehicle on video tape and see what you were doing wrong. Also if you are a beginner you can watch these tapes in your spare time and you will learn very quickly. This set comes with one basic car opening video that shows you how to use each tool in the set. The other nine videos are available separately at a cost of around $50 each. This set also comes with a hard case that holds all of the tools in the set.

We are not aware of any other car opening tool company that has videotaped each and every car opening in their manual in this way. The only other car opening videos that we are aware of cover only a few fundamental opening techniques. Alternate methods of opening are also included in this manual. If you need to open a car by picking the lock, the direction of turn to unlock the car is also given for each make of vehicle.

Yearly updates for the convenience manuals are FREE. If you would like to update the Quick Entry Manual on this set, the cost is about $25. This will completely update your manual including new methods, new alternate methods, and opening the latest vehicles. If additional tools are required to open new vehicles, they will let you know at this time and it will be at your option to purchase them.

This manual gives line drawings instead of photographs for illustration purposes. We feel line drawings are better than photographs which can be fuzzy and hard to understand. The manual also gives warnings in bold print to alert you to potential problems on particular vehicles.

An added bonus for purchasing this set is that you will automatically be added to their mailing list to receive their newsletters. This newsletter is very informative and will help to keep you up-to-date on new vehicles.

Here is the address and ordering information:

Tech-Train Productions
P.O. Box 15401
Pensacola, FL 32514
Order by calling:　1-800-356-0136

High Tech Tools Model 2000

This company is our second choice for car opening tools only because they do not offer the car opening videos that we feel are very beneficial for a beginner. They do include a basic car opening video that covers basic opening techniques with each tool set you order.

This is a very good car opening tool company that has an easy-to-understand opening manual using line drawings for illustration purposes. They also provide alternate opening methods and update their manuals every year. There is also a hard case for your tools available.

This company does have some advantages over the others if you are interested in more than just opening cars. With your car opening tool set they provide a two-volume encyclopedia (over 1500 pages) of automotive lock servicing information such as: steering column servicing, air bag servicing, V.A.T.S. servicing, domestic ignition locks, part numbers, key code locations, impressioning methods, and more.

The yearly updates for the manual on this set run about $90 and include any new tools that may be required to open the newer vehicles.

Here is the address and ordering information:

High Tech Tools
Drawer 450370
Miami, FL 33145
Order by calling: 1-800-323-8324

Grand Master Z-Tool System

This set is sold by Slide Lock Tool Co., Inc. Many locksmiths that we know personally prefer this set of car opening tools because there are fewer tools in this set and the opening manual is a smaller size. Although there are fewer tools, this does not affect the number of cars you can open with this set. The manual is easy to understand and uses drawings instead of photographs for illustration purposes. A hard case is available to hold all your tools and the manual is updated every year.

The yearly updates for the manual run around $40 and you do have the option to purchase any new tools that may be required to open the latest vehicles.

Some of the opening techniques in this manual require you to bend your tools to match a drawing in the manual. For this reason we don't feel comfortable recommending these tools to a beginner, although this is a very good set of tools.

There are no videos associated with this car opening system but this set will open virtually any car on the road.

Here is the address and ordering information:

Slide Lock Tool Co., Inc.
1166 Topside Rd.
Knoxville, TN 37777
Order by calling: 1-800-336-8812

 **Tool List**

This is the list we promised you at the beginning of this section. If you order these tools, you will have the very best tools available to you. This is very important to your success.

1. *Tech-Train 2050 Master Car Opening Kit*
2. *A Lock Picking Book*
3. *A Lock Picking Video*
4. *Deluxe Lock Pick Set* (16 tools w/leather case)
5. *A pick gun*

Keep in mind that the only thing you really need is the car opening kit. Everything else is an optional backup system for you.

Training and Education

SECTION OUTLINE

SECTION 6

Training and Education

Introduction

Locksmith training, unless you enroll in a campus course (full-time day classes) which very few of us starting out can do, is basically a self-educating process. Even the formally trained will become self-educating after graduation because the training never ends. You do not need formal training to open cars but car opening classes and seminars are available and may be helpful. New automobile designs change. New cars sometimes require new opening procedures and a new tool is invented to open them. Some car manufacturers try to design locking systems that cannot be bypassed in an effort to keep the cars from being stolen. If you subscribe to a trade magazine, you should know about these changes and how to deal with them before the cars are available for sale. This is how we all keep up with the never ending changes of this trade. It is part of staying informed and being up-to-date.

Your training begins with the tools you order. Study the books and the instructions that come with your tools. Memorize the names of your tools. Study your car opening manuals and memorize the opening procedure for each different vehicle. Remember which tool is used for each vehicle. This will take some time but this is how you become a car opening expert. You can always look up a vehicle in your opening manual at the jobsite (if you're a sissy).

Continued education in locksmithing is accomplished in very much the same way that doctors or lawyers continue their education. It is done through books, videos, newsletters, magazines, seminars, supplier-sponsored classes, trade associations, trade shows, schools, correspondence courses, and relationships with other people in your trade.

In this section we will take a look at each individual educational tool at our disposal.

 **Experience and Practice**

O.J.T., or on-the-job-training, under an experienced employer is the very best training you can get but it comes at a high price. You work long hours, nights, and weekends at minimum wage. It also costs you your independence; you are working when someone else wants you to. The only skill that you need to be successful in car opening is the ability to read and follow instructions. Most of the time there is a picture associated with the instructions so, after a short time, you won't even have to read the instructions. You will have learned to recognize what the picture is telling you to do!

When you are starting out you will need to open as many cars as you can get your hands on until you develop the knack or "feel" for what you are doing. It actually is a feel that you are trying to develop, what some people call being able to "see with your hands."

Getting the practice you need is easy. Start with your own vehicle and then do your friends' and neighbors' cars and pickups. After that stop by a small used car lot (they will be less intimidating when you are first starting out), introduce yourself, and tell them that you are in the car opening business. Say that you spotted a couple of cars on the lot that you have not opened before. Offer to give them your card for a free opening whenever they need it, if they will let you practice on a few cars on the lot. We do this ourselves when training a new employee and we have never been turned down. Occasionally we gain a regular customer!

Do this over and over again until you lose your apprehension about opening cars for paying customers. Soon you will be a fearless self-proclaimed car opening expert!

 Books

Unlimited books are available to you on every subject you can imagine in this field from *Lock Picking*, to *Service with a Smile*, to *Safe Cracking*, and more! The only thing you need to know about books on various subjects in this trade is where to get them. When you subscribe to a trade magazine you will find a list or two that you can order from right in the magazine. Your supplier will have a list of books for sale also. Some trade associations have a free library of books and videos available to members. These have usually been donated by other members.

 Videos

It's A Fact!

There is no substitute for "hands-on" experience.

Videos are fairly new to this trade but the number and kinds available are increasing. These are normally sold in the same location in magazines as the book listings. Car opening videos are available with each make and model vehicle on the road shown with the door panel removed so you can see exactly what is happening inside the door. Watching a video is just like having someone standing beside you showing you how to open the lock. You will get a much clearer understanding of the subject being taught by watching a video than by reading a book. The video may be more expensive but we recommend trying to get the video rather than the book, if the option is available. This form of training is very effective but remember that anything you are learning should be considered theory until you have actually done the work yourself. There is no substitute for "hands-on" experience.

 # Magazines

Trade magazines are one of your least expensive forms of education and an excellent method of keeping up to date on new products, tools, service procedures, and techniques. When you subscribe to one of these magazines you'll get one issue every month and most of them offer subscribers a bond at a very low price. Start with one magazine subscription and read each issue cover to cover. One magazine is all we have time to read but if you find that you have the time to read more, then subscribe to another one.

Most locksmiths are bonded and insured and many customers look for this in your advertising. Most people think that in order to get a bond you have had to go through an extensive background check and that only a select few who can prove their honesty and trustworthiness can get a bond. This is good for your business. A bond is simply insurance for your customer that guarantees that money is available to reimburse your customer should you be convicted of stealing from him.

It is very easy to get a bond. Fill out the order form that is in the magazine you've subscribed to, send in your payment, and in about three weeks your bond certificate will arrive in the mail. There are no background checks. Nothing is required other than being either a student of locksmithing or a practicing locksmith and a subscriber to the magazine. We are not trying to belittle the "Honor" associated with the bond. If you are not an honorable person you should not be in this trade and will not be working long before you get a very long vacation, all expenses paid (Incarceration)! If you are a locksmith, these vacations are longer than they normally would be.

Each magazine has its own bond. Bonds vary from a low of $5,000 to the highest we know of—$15,000 through *The National Locksmith Magazine*. Your cost is very low, $5 to $15 per year, and the bond normally expires with your magazine subscription. Ask the other locksmiths in your town which magazines they subscribe to. We have listed our personal favorites here in order of our preference. Call these magazines and ask for a free copy. This is how they try to get you to subscribe anyway. They will send you a couple of free copies and ask you to subscribe. They are different, so this is a good way to decide which one you like the best.

The National Locksmith
1533 Burgundy Parkway
Streamwood, IL 60107
Phone: 630-837-2044

Locksmith Ledger
850 Busse Hwy.
Park Ridge, IL 60068
Phone: 847-692-5940

Schools

We are giving you a list of schools in this Manual just in case you are interested in getting some classroom instruction. We are unable to give costs for instruction at each school but you can expect to pay from $2,000 to $10,000 (and higher) for classroom instruction. These costs vary greatly from school to school and some schools prefer to customize classes to meet your needs. This is much more cost effective for you. You may be required to take a test (for a fee) before you enroll to see where your weaknesses are and you'll receive counseling as to which areas need improvement. Most schools want to make sure you actually know what you think you know before training you in other areas. Look over the chart in the section on **Schools**. Some are correspondence courses and some offer classroom instruction. We have also included estimated completion times for some of these courses.

Correspondence Courses

These are locksmith courses that allow you to study at home. Individual lessons are sent to you; when you complete them you send them back to be graded and they are returned to you. All of this happens through the mail, so don't rely on the estimated completion times to be very accurate. You can request that more than one lesson be sent to you at a time and this will speed things up quite a bit. There is also a phone number you can call, if you've signed up for the course, so you can speak with an instructor if there is something that you just don't understand in the lessons. You will receive a certificate upon completion that states that you are now a "certified locksmith" from their school.

A list of these schools is mixed in with the schools list; refer to the school chart to find out which ones are correspondence courses or home study courses. These courses can be done at your own pace and are much less expensive than classroom study courses. Correspondence courses cost anywhere from $500 to $2,000. Almost all of them offer a payment plan that is very affordable and which makes these courses available to almost everyone. The completion times here are suggested times from each school but you can complete a twelve-month course in three months if you want to study hard.

We discovered **how to save up to $350 on correspondence courses** quite by accident. This held true for every one of them that we checked into. We were looking for a better way to train employees that would not take up as much of our time as showing each one of them every little trick and service procedure that we've learned.

It's A Fact!

You can turn $50 into $350 if you can wait long enough.

We called several correspondence schools and asked for information on the courses that they were offering. We were surprised at how quickly we received the information from everyone. We looked over the courses to see what and how in-depth the subject material was. Some were better than others. We were very busy so we set the material aside and did not respond to any of them. About two weeks later we began receiving coupons in the mail from these schools offering $50 off if we would sign up right now, saying we'd better hurry because this offer expired in two weeks. We were excited but still very busy, so we did not respond. Just like they said, that offer expired. But they sent us another offer, a coupon for $100. Now we did not have time to waste; we only had ten days to respond!

This could be a long story so we'll cut it short. Next came a coupon for $150, then $200, then $250, $300 and finally $350. The interesting thing about this is that as the coupons became larger, the time between the new offers became longer, and the coupons expired sooner. It took eight months from the time we called for information to get the last coupon for $350 with six weeks time elapsing between it and the earlier $300 coupon. Four months after our $350 coupon expired one of our customers asked how he should go about becoming a locksmith and we recommended one of these courses to him. The school accepted the $350 coupon we gave to him that had expired four months earlier! If you can wait, you can get some of the courses for about half price!

 # Trade Associations

What is a trade association? A locksmith trade association is a group of individual locksmiths who have come together with a common desire to make the locksmith industry better. Better for the community. Better for the locksmiths. Better for the industry. They want to educate themselves, to make more money, and to make the trade more fun.

There is normally a yearly membership fee for joining the association which helps cover the cost of classes, the meeting room, supplies, newsletters, and so on. The association does not collect fees for profit and you will more than get your money's worth in education alone. This is also a great place to get referrals and help from other experienced locksmiths. The main thrust of most associations is to develop a "let's work together" attitude. Some of the locksmiths will only refer business to other members of the association in an effort to get more members to join.

When you join one of these associations you will be expected to adopt their Code of Ethics which basically is a "Code of Honor." Locksmithing is an old trade; its secrets have always been guarded. You should be proud to be in this trade. Your customers must know they can trust in your honesty completely.

Make as many friends as you can in this trade. You'll find that you can't have too many friends in this business. One very important thing to remember is not to undercut prices. Ask the other locksmiths what they are charging and follow suit. You don't want to get customers because you are cheaper than everyone else. This will only cost you money in the long run because you'll lose referrals from the other locksmiths.

> **NOTE:** You can actually build a very good business on referrals from other locksmiths. It is essential that you establish a good working relationship with others in your trade.

If you are undercutting prices, no one will be willing to give you a hand should the need arise, and believe me—it will. The old song "Everybody Needs Somebody Sometime" holds true in this trade. You should get more out of joining an association than anything else available to you.

The list of associations we have provided for you is not complete. New groups are being formed daily. Before you contact any of the groups on this list, check with the locksmiths in your area to find out if there is a local group in your town. Also find out which associations the locksmiths in your area belong to. Be interested and ask questions!

Associations are not normally considered businesses, therefore most of them do not have phone numbers. Don't be surprised if you call and someone answers with a business name other than what you expected. Officials are normally elected once a year or every six months from the members who are probably operating locksmith businesses during the day. Write letters requesting information on the groups you are interested in. Ask to be added to their mailing lists in order to receive newsletters from their associations.

Newsletters

Occasionally you will receive newsletters from your suppliers letting you know about a new product or service they are offering or changes in company policy. They will also keep you informed on upcoming seminars and trade shows they are sponsoring. Trade associations send out newsletters once a month to everyone on their mailing list and most do not require that you join their association to receive their newsletter.

These letters are full of all kinds of information. The best thing that we personally like about the association newsletters is that they keep us informed, several months in advance, about all the seminars and trade shows that are coming up. There is usually a good article on a service procedure, a new trick, or a tip of the month. Also, we find these newsletters to be more concerned with local issues that affect our businesses, such as a new law in our town or local police and fire department policies regarding car openings and deadbolt installations.

 ## Seminars

Seminars come in two forms—classes or lectures. They can be as short as two hours or as long as two or three days. Usually a seminar is a short three- to four-hour "hands-on" training class that your supplier has arranged to hold at his location during the evening (after hours between 6 p.m. and 10 p.m.). These lessons are on the disassembly, repair, rekeying, installation, bypass methods, or selling strategies of one product or product line. There is a fee to attend these classes but you usually get more free product for attending than you could have purchased for the cost of the seminar. Occasionally a supplier will put together a weekend seminar that lasts all day Saturday and all day Sunday and offers a selection of classes for you to choose from. These two-day seminars are held at local hotels and provide low-cost training you should take advantage of. These seminars are an excellent way for you to get to know your suppliers. You will be notified by your suppliers and through your local association newsletters of these upcoming seminars.

 ## Trade Shows

A trade show is something like a locksmith carnival. A trade show is a chance for manufacturers to show their products and try to sell you on the idea of selling their products versus their competitors'. They will do many different things to entice you to come to the shows such as giving away door prizes and having free drawings for prizes during the show. These are sometimes very good prizes worth from $5 to over $1000. It depends on the size of the show. Some offer prizes or samples just for visiting their booths. Lunch is almost always a lavish buffet and free to everyone. Lockpicking, safe opening, and impressioning are some of the "compete for prizes" games you'll find here—all at no cost to the participants. Manufacturers, tool companies, alarm companies, dog trainers, everything that you can imagine that is even vaguely security-related, can be seen at the larger shows. We have even seen spy products on display with factory representatives on hand to show you how everything works. We've seen cameras installed in car radio antennas so you can watch people without looking at them, all kinds of listening devices, (including wiretapping systems), and a small machine you hook up to a safe dial called an auto dialer that will automatically dial every possible combination in less than 24 hours until the safe opens. Some of this stuff will really blow your mind! You'll walk away saying, "Man, I thought that stuff was only in the movies!"

Manufacturers set up booths so that you can walk around and look over their products and ask questions. You can be sure that there is probably no one who knows more about these products than the manufacturer's representatives. Some locksmiths (ourselves included) go to these shows, large or small, just to hang around and get to talk to some of these experts. You will be surprised at the questions you'll come up with when you're talking with an expert.

We've always come home with enough free stuff to at least cover the expenses involved in getting to the shows. We recommend you attend these shows whenever possible, not just for the free stuff, but because you will learn a great deal about the products you are using and discover new products. This knowledge will enable you to make more money or make money faster by using a new tool or procedure you learned about at the show.

Most of the smaller trade shows are put on by suppliers. Your supplier will invite manufacturers of the products that they sell to come and answer questions. The manufacturers will display products that they would like to sell more of. There are many reasons suppliers are willing to put on these shows. One is to promote a relationship of "we are here to help you" between you and your supplier. If you have a good relationship with a supplier, you will continue to order from him. The discounts offered at trade shows, large or small, cannot be beat and many locksmiths will wait to order products or tools if they know a show is coming soon. You can get up to 50% or more off products and tools at the show, but only if you order at the show. This discount is usually not limited to items on display; it includes your supplier's entire product line!

Most of the larger shows are put on by associations. One of the largest associations is ALOA which stands for Associated Locksmiths of America. This group puts on a show once a year and is the biggest show that we have ever seen. It could take you three days just to walk around and look at every booth. It is held in a different town every year and usually lasts for at least a week. During this time classes are available all week long and the instructors are experts in their fields. You will find out when and where this show will be through the trade magazine you've subscribed to. Usually you will read about it at least three months in advance. When you see this show being advertised, you will need to call the advertised number to have ALOA send you an information packet which includes class information, discount airfare, hotel arrangements, and so on. If you decided you were only going to go to one show in your lifetime, this would be the one!

Relationships/Networking

If you are seriously interested in learning other areas of locksmithing, one of the best ways to learn is from someone who already knows and is willing to teach you. The trick is getting him to teach you. Most people that you ask to show you how to do something will ask themselves, "Why should I?" and "What's in it for me?" Of course, they will not come right out and ask you these questions. Instead they will lead you on with, "Sure, when I can find the time." They may simply refuse with the excuse, "I just don't have the time to train you."

Developing a good relationship with a few locksmiths in town is a good place to start. If you approach them properly, you can get them to train you. We see this happening every day in our town. A beginner shows up in town and wants to learn everything there is to know about locksmithing. He begins making friends with the other locksmiths and ends up being trained by not one but usually three or four different locksmiths. Your approach must be handled in the following way to be effective:

1. Be honest, trustworthy, and dependable.
2. Be genuine. Make him feel like he is the very best in his field.
3. Validate that person's value as a professional.
4. Never ask or assume that anything is free.
5. Always offer to pay for his time and materials.
6. Tell that person that you would feel privileged to be taught by him.

If you go about asking in this way, we don't know many people who would turn you down. Some will train you free if they know you consider them important and that you really care about learning from the best and doing the job right.

There are other relationships you will need to develop to be super-successful in this business, but they do not fit into this training and education section. You will find them in the section on **Getting Business**.

☞ It's A Fact!

If you approach someone properly, you can get him to train you.

Paperwork

SECTION 7

Paperwork

Introduction

There are only a few forms that you will need to use in this business. We are covering the basic forms that you will be using to run your business (excluding any forms used for bookkeeping) in this section. You may find that these forms are the only forms you will need to use. These are all of the forms that we use.

 ## Work Tickets

Every time the phone rings you should be reaching for a work ticket at the same time you are answering the call. A work ticket is used to record the information you need to be able to go out and do the work. It can be a customized form that you have made at a printshop, a legal pad, a post-it note, or a piece of scratch paper. Always have a pen and paper on you for this purpose.

It almost goes without saying that you need to do this. We are covering the work ticket here because there is some specific information that you will need to collect that will make your job easier.

You should get as much of the following information as you can on your work ticket:

1. The name of the caller
2. The person to contact at the jobsite
3. A phone number
4. Jobsite address/directions
5. Vehicle description—year, make, model, color, license plate number
6. Quote given
7. Method of payment
8. Estimated time of arrival given to customer
9. Time the call ended.

Do not turn down a job just because you cannot get all the information that you want. You will be surprised at the number of people who are not going to know exactly where they are, their telephone number, their address, apartment number, or even what kind of car they are driving. They usually know what color it is, but not always (no kidding). The more information you can get, the faster you will be able to complete your job and get on to the next one.

Many times a convenience store employee, hotel manager, shopkeeper, or Good Samaritan will call for your customer. When this happens, your customer may not feel responsible for your fee if they did not personally request the service. Always ask to speak directly to the customer and have them agree to your fee before you accept a job.

Name of Caller and/or Person to Contact at the Jobsite

Your customers will only give you the information you ask for. You may arrive at the address they give you without knowing that you were going to find a large office building or apartment complex. If you have only asked for the address, that is all you will get—the building address—and you don't want to have to spend thirty minutes trying to find your customer. You would expect that, if it was a large office building, your customer would tell you to go to the front desk and ask for them by name. If it was an apartment complex, you would expect them to give you the apartment number. The customer is upset and they will not remember to give you important information if you do not ask. Your customer will be waiting in an office or apartment for you to arrive and you will not know how to find them.

Method of Payment

Don't be afraid to ask what method of payment the customer will be using. They may be planning to pay you with a postdated check, out-of-state check, or a credit card that you are not authorized to accept. We have opened many cars (before we adopted this policy) and had our customer say, "Okay, now you will have to follow me down to the ATM machine so I can pay you." This usually takes another thirty minutes of your time and the customer usually does not want to pay you for the time you are losing. Time is money in this business. Always ask what method of payment they will be using before you go to the jobsite. If they are going to want you to follow them to an ATM machine after you have opened the car, you will be able to quote a higher price for this service ahead of time.

Time the Call Ended

If you are running two or three lockouts at the same time you will want to know how long each customer has been waiting. A glance at the "Time the Call Ended" tells you which customer is next, if you are on schedule, and helps you plan how to handle any new calls.

 # Authorization/Release Form

An authorization/release form (see sample below) protects you in two ways. The first item, the authorization, protects you from having criminal charges attached to you in the event that you have opened a car for a thief or any unauthorized person. When you require your customer to show a driver's license and car registration on each car opening, you can never be charged with negligence in the conduct of your business. You are, for all practical purposes, exercising "Due Care." Never complete a car opening without getting this identification. You will not be able to check the registration until you open the vehicle. If the registration does not match the driver's license after you have opened the car, it will be at your discretion whether or not to call a police officer and file a report.

WARNING: You must have the customer fill out the authorization/release form and sign it—before you open the car.

| INSERT YOUR COMPANY LOGO, ADDRESS AND PHONE NUMBER HERE | **AUTHORIZATION/RELEASE**

I legally can and do authorize the work described below and hold bearer harmless from any resulting damage or claims. |

Name: _____

Address: _____

City: _____   State: _____   ZIP: _____

Phone: _____   DL#: _____

Description of Work: _____

_____

_____

Year: _____ Make: _____   Model: _____ Lic. Plate #: _____

Method of Payment: ❑ Cash   ❑ Visa   ❑ Mastercard   Other: _____

_____   _____
Customer Signature   Date

Always
fill out
the
forms

The second item, the release, is a damage waiver that protects you from having to repair any damages that you did not cause. If you damage anything, you need to be responsible for it. On the other hand, if somebody tried to open that car before you got there, then you need to make sure that your customer understands that you're not going to be responsible for any damage done before you arrived. There have been many times that we have rolled up to a job and found clothes hangers sticking out of the window, paint damaged because of an attempt to wedge the door open, scratches on the window tinting, or broken molding.

There have been times when a security guard or police officer has tried a Slim Jim tool on a car that does not have any place for the Slim Jim to work. Someone trying to help may tear up electrical wiring, break plastic parts, bend linkage, scratch paint, or break the molding. If the customer does not notice the damage at that time, you may look guilty if you were the last person working on the car. So if there is any kind of damage anywhere on the car, you need to have the customer sign the release form first, and then you can open up the car. Explain to them, that if somebody worked on the car before you, you are not responsible for any damage they may have done.

This will give you an idea of the repair cost when someone does damage a door by attempting an amateur opening. We estimate about $125 (plus parts) per door to repair the door. If you're going out to unlock a car for $30, you don't want to spend $125 to fix the car because you forgot to have the customer sign a release form. That sure makes for a lousy day. Plus—that doesn't include the time that you will lose to tear down a door panel and put it back together correctly.

You can see why it's very important for you to have an authorization/release form, and that release form needs to be signed by every single customer. The one time that you don't have the customer sign it will be the time the customer calls you back and says, "You did the damage, and I want you to fix it." There's not a thing you can do about it. If you are taken to court, the court is going to say, "Sure, somebody else can fix it, but you're the one who will be responsible for paying for the repairs." That will come out of your pocket—not that of the person who did the damage or the customer's—but your pocket.

Where Do You Get Authorization/Release Forms?

These can be ordered from a locksmith supplier or you can have your own forms custom-made at a printshop using the sample on the previous page. You can save money by ordering pre-made forms but you will eventually want to have this form printed directly on your invoices to cut down on the amount of paper you are handling.

Invoices

You can probably find a suitable blank invoice at an office supply store to use for your car opening business. This will be the cheapest way to start. As your business grows you should have your invoices custom-made with your company name and logo on them to improve your professional image. The following checklist covers the important information you need to include on your invoices. The sample invoice on page 66 can be taken directly to your printshop. The printer can insert your logo and company information.

What Information Should I Have on My Invoices?

☐ 1. Your company name

☐ 2. Your mailing address

☐ 3. Your phone number

☐ 4. Logo (if you have one)

☐ 5. Invoice number

☐ 6. Date

☐ 7. Customer's name, address, zip code, phone number

☐ 8. Job location

☐ 9. Work ordered by

☐ 10. Work performed by

☐ 11. Terms

☐ 12. Due date

☐ 13. Description of work

☐ 14. Authorization/Release

☐ 15. Purchase order number

☐ 16. Method of payment

☐ 17. Quantity

☐ 18. Unit price

☐ 19. Labor

☐ 20. Materials

☐ 21. Tax

☐ 22. Total

☐ 23. Signature line for—"I acknowledge satisfactory completion of work."

☐ 24. Returned check fee

☐ 25. Past due account fee

☐ 26. Guarantee

☐ 27. Thank you

INVOICE #:

INSERT YOUR COMPANY LOGO, ADDRESS AND PHONE NUMBER HERE

Date: _____

AUTHORIZATION/RELEASE

D.L. # _____	D.O.B. _____
WT. _____	HT. _____
HAIR _____	EYES _____
AUTO:	Year _____
Make _____	Model _____
Lic. # _____	

NAME _____

ADDRESS _____

CITY _____ STATE _____ ZIP _____

JOB LOCATION _____

PHONE _____ P.O. # _____

X _____

WORK ORDERED BY:	WORK PERFORMED BY:	**TERMS:** ❑ Cash ❑ C.O.D. ❑ Credit Card: _____ ❑ Other: _____

DESCRIPTION OF WORK	NO.	AMOUNT	MATERIALS	LABOR

I acknowledge satisfactory completion of work.

X _____

PLEASE PAY FROM THIS INVOICE

$25 charge for returned checks.   **Total amount due by:**

Total Labor	
Total Materials	
Tax	
TOTAL	

A service charge of 2% per month which is 24% per year will be charged on past due accounts over 30 days. The cost of collection will be added to amount due.

GUARANTEE: Workmanship performed and materials installed are warranted for ninety days. During this period, if trouble develops in work performed, replacement of defective parts will be made free of charge.

Thank You!

Yes, you are going to spend more time filling out your invoice than it is going to take to open the car, but getting complete billing information is very important, especially with your regular accounts.

What Is Complete Billing Information?

Complete billing information means that some companies require a purchase order number or other information found on your invoice. If you don't have a purchase order number on your invoice, or any other information requested, the company won't pay you. You may have to put a job address and a purchase order number on your invoice for a corporate customer. It all depends on the company that you're working for. If you're working for Emergency Road Service, sometimes you're required to get a membership number, a purchase order number, the VIN number (vehicle identification number), and the correct billing address. If you write down one wrong item on this invoice and everything else is correct, the company won't pay you. It is extremely important to get complete billing information on your invoices.

Are Invoice Numbers Important?

Yes, and if you have a place for all of the other information on your invoice, then you won't miss any vital information. You may need that information in the future. If you have to call a customer back, you will know who to talk to. Invoice numbers will help you keep track of each lockout you accept and the IRS prefers that you keep track of each and every invoice. If one of your invoices is missing, the IRS may charge you taxes and penalties for the amount that the IRS will estimate was on that missing invoice, so keep good records to avoid the penalties and inconvenience.

How Do You Give a Description of the Work?

Give a description of the work that you have performed. Sometimes it might be as simple as writing down "lockout, 8:00 p.m." and the type of vehicle (Chevy Corsica, 1988). If there was something damaged on the car when you arrived, write down what was damaged on your invoice. The customer may call you back and say, "You damaged this piece of molding on my car," or "My lock doesn't work since you opened my car." If you have written on your invoice that the lock was already broken and the customer signed the bottom of that invoice where it says "I acknowledge satisfactory completion of work," you have protected yourself from having to repair damage that you did not cause. Be sure that you get both a description of the vehicle and any problems you had noticed, written on your invoice. Some examples of what to write are: the window tinting was scratched, paint was damaged, doors did not operate correctly, lock was broken, key jammed in lock. The extra time you take to write these things down may save you time and money in the future.

 # Statements

Statements provide a summary of the work that you have done for customers who have charge accounts with you. Most of the time when you do lockout work, you will be getting your money right away. But if you get into the emergency road service work like AAA or Chevy Roadside Service, or if you work with car dealerships, you will not be getting paid right on the spot. These companies will insist on having charge accounts with you. This means you will be getting paid every 30 days (or sometimes 60 days) after the job is completed. For example, a roadside service company will call you and say: "I've got a job in your town. Do you want it?" The company will give you all the information you need to put on your invoice and when you finish the job you will have the customer (the person you opened the car for) sign the invoice.

There are a lot of little Emergency Roadside Service (ERS) companies that you may only do one or two jobs a month for. You can send those statements in right away. There are other roadside services that call us to do five or ten jobs a week. They don't want you to send them a bill every time you do a job. They want you to hold your invoices and send one bill at the end of the month. The roadside service company will pay all of the invoices listed on that one bill. Instead of getting fifteen or twenty checks for $30 you will get one check for $600. The ERS companies may require a copy of each invoice and they will want a bill that summarizes everything you have done for that month. When you work with these types of companies you may have to have a purchase order number on every invoice.

How Do You Write a Statement?

This is easy to do. Most of the time all that you have to write on your statement is an itemized list of your invoices. For each invoice you will list the following things:

1. The date the work was performed
2. The invoice number
3. The amount of the invoice.

Then you will total all the invoice amounts. We recommend, and some companies will require, that you make a copy of each invoice and send these copies with your statements. You will simply staple your statement on top of your stack of invoice copies and mail it. We are still using blank statement forms that we purchase at an office supply store. These are very inexpensive and there are many different styles to choose from.

A sample statement form is shown on the next page.

STATEMENT

28663

Date: _____ 1-1 _ 19 99 _

TO _____ Your customer's name _____

_____ Your customer's address _____

_____

Terms: _ Due _____

IN ACCOUNT WITH

[INSERT YOUR COMPANY NAME AND ADDRESS HERE]

DATE		INVOICE #				
1	12	0001			35	00
1	15	0002			35	00
2	24	0003			35	00
3	06	0004			35	00
		Total			140	00

# Contracts

Contracts are not a form that you will use to run your business. However, since many people believe that we use contracts all over the place, we thought that this was the best section to discuss them.

You are going to hear other locksmiths saying, "I have this contract," and "I have that contract." In reality they may have no contracts at all. They are just telling you who their regular customers are. There are very few contracts to be had in this business.

If you tried to present a contract to a business they would probably laugh you off the property. A contract is a one-way street: your customer's way. When you're dealing with contracts, the contract won't be from you to your customer; it will be just the opposite. The reason for a contract is because your customer wants a discounted price or a set price that you agree not to exceed. If they have a lot of work for you, they will want you to sign a contract saying that you will not charge them any more money than the amount on their contract. This agreement could be in the form of a lockout service account for an ERS company or an automobile dealership. If these bigger companies have enough business for you, it's worth signing their contract to get the repeat work. Be warned. This means you may have to do after-hours calls for them at your normal daytime rates.

Contracts are a good way to assure repeat business. You will not have a contract that you set up and the ERS signs for you. It just doesn't work that way. Be careful what you sign. You don't want to get into a position where you can't make good money for your labor. Always read a contract completely before you sign. These contracts should be absolutely nothing more than an agreement on price. If you ever come across a contract that is anything more than an agreement on price, don't sign it. Remember, if you get into a situation where you are not making money, simply tell your customer that you can no longer honor the contract and that they will have to use another lockout service. You can refuse service to anyone and this will cancel your contract.

Money Matters

SECTION 8

Money Matters

Introduction

This section covers all the different forms of payment you may be asked to accept and how to handle them. You will learn how to properly accept a check and how to collect on a bad check. You will get familiar with accepting charge cards, setting up and handling charge accounts, and using purchase orders. We will also show you how to set your pricing structure and determine how much you will charge for your services.

How Much?

The first area we are going to cover is how much to charge for your service. You can actually charge whatever amount of money you can get someone to pay. There are several ways to figure out how much you should charge for your service. The best way to start is to find out what the going rate is for opening cars in your area and then adjust up or down from there.

Do this by calling your competition and asking them how much they are charging to open cars. Your two main competitors are going to be locksmiths and tow truck drivers. If you let them know that you are the competition calling to check prices, you are not going to get accurate information. They may give you high prices in an effort to make you the highest-priced car opening company in town or they may give you ridiculously low prices to discourage you.

Here's an example of how to make these calls. Go through the yellow pages and call up a locksmith. Pretend that you have locked your keys in your car. Before they give you a price they are going to want to know what kind of car it is and the location of the vehicle. Give them all the information they want. Find out how much they charge and then say, "I'm going to call a couple more places. Thank you." It is important that you say this to avoid confusion and to let them know that you are not asking them to send someone out.

This should be done for every single locksmith and tow truck driver in the telephone book. It will give you a very good idea of what the prices are for opening cars in your area. You should not feel bad about doing this. Competitors price check each other all the time just like you are doing now. You will need to call each company that you are price checking at least twice (at different times of the day and night), because prices change depending on the time of day and day of the week.

An example of our prices is as follows:

$25	9 a.m.– 6 p.m.	Monday through Friday
$35	9 a.m.– 6 p.m.	Saturday and Sunday
$35	6 p.m.–10 p.m.	
$45	10 p.m.–12 a.m.	
$65	12 a.m.– 6 a.m.	
$35	6 a.m.– 9 a.m.	

Don't be concerned with our prices being too high or too low. You will need to find out what the going rates are for your area. The average price to open a car in our area during the day is $30 but if you were to drive 45 miles to the north of us, the average daytime rate in that area is $65.

You will want to set your pricing so that you are not the cheapest or the most expensive. Pick a price somewhere in the middle. It seems that the people who have the lowest prices end up getting a very high percentage of bad checks.

■ **Reminder**

- **Charge a reasonable fee.**
- **Be competitive.**
- **Do not give away your service.**

There are other circumstances that affect pricing such as weather, special occasions, and holidays. At these times your prices should go up. We may charge as much as $100 to open a car if the weather is extreme or if it is a special holiday like Christmas or Thanksgiving. If you are not sure what to charge during these times, just call your competition and ask how much they are charging. Pay particular attention to how many of them are not even answering their phones. If you are one of the very few services available, because others are not answering their phones, then you should be able to charge more for your service. We have had customers tell us that they were charged $200 to have their cars opened on Thanksgiving or Christmas Day. We could not sleep at night if we charged these prices, but you need to be aware of supply and demand. Do not give away your service.

Most of the time when a customer calls you can quote a flat fee. There are times, however, when you will need to consider mileage. If someone is calling outside your normal service area, consider charging your flat fee plus $1.00 for each mile outside your normal service area. If you live in a rural area, everything may be based on mileage instead of a flat fee. This would depend on the area in which your business is located. Find out what your competition is doing and follow suit.

Another way to charge for your service is to charge a service call and an additional fee for the type of car you are opening. A good example is: $20 for the service call plus (if it is a car that is very easy to open) an opening fee of $5 or $10. If the car is difficult to open you might charge $15 to $20 in addition to your service call. We do not recommend charging your fees in this way because there will be times when you arrive at a job and find that the customer has already solved the problem. Naturally, you will want to collect the full amount that you quoted over the phone instead of for just a service call. After all, your time was spent driving to and from the job; it only takes a few seconds to open the car. It will help you collect your fee in this situation if, when they ask "How much to open my car?" you answer that the charge for *coming out* is $$$.

Cash

Cash is always the safest kind of payment to accept. There is no risk involved. All of the other forms of payment that you will receive are actually promises to pay. Eventually everything is converted into cash (if you're lucky). If you get paid in cash, then everything is done. Give your customer a receipt and the job is finished.

Checks

Checks are the most common form of payment you will be asked to accept. You are not required to accept anyone's check as payment for your service. If you accept checks you are going to get some bad checks that you will not be able to collect. However, you will lose much more business by not accepting checks than you will lose on those occasional bad checks. We recommend that you do accept checks as a form of payment. This is a decision that you will have to make about running your business.

 Guidelines for Accepting Checks   There are several guidelines you need to follow for accepting checks. These recommendations will help you decide which checks to accept, which ones to reject, and how to collect on the occasional bad check you have accepted.

Be Cautious of New Accounts

A very high percentage of the checks returned for insufficient funds are written on accounts that are less than one year old. The check number appears in the upper right-hand corner of all checks. Be very careful of checks numbered 101 to 200.

Many banks are now printing a date code on checks. This code usually appears near the address information. The date code is a three- or four-digit number such as "1297" which indicates that the account was opened in December 1997.

Require Two Forms of Identification

The requirement of a valid driver's license or an official state identification card is an absolute must for accepting checks. Write the license number on the face of the check along with any other information that is not printed on the check such as address and both home and work telephone numbers. Make sure that the license is valid, check that the person offering the check is the same person as in the photograph, and compare the signatures for a good match.

As second identification, the "Guaranteed Check Card" is preferable. Be sure to comply with the provisions mandated by the issuing bank. You will find these on the back of the card. Unless these procedures are followed, guarantee of the check is not assured by the bank.

If the customer does not have a "Guaranteed Check Card" (actually, most people do not have one), ask for a credit card from a local department store. National credit card companies will not share information about their customers. Local stores are usually willing to give out information which will aid in the apprehension of bad check writers. Check the signature on the card and do not accept any card where there is no signature.

Place All Information on the Front of the Check

Writing information on the back of the check is useless. This information will likely be covered with bank and clearinghouse stamps. If the check is processed for prosecution and any part of the key information is unreadable, prosecution is very unlikely. Use the cross method, but be careful not to write over any of the printing on the face of the check. Example:

Driver's License Number	Credit Card Number
Salesperson's Initials	Other I.D. or Phone Numbers

Always Initial Checks

Always initial the checks that you receive. Place your initials on the face of the check with the other identification information. If the check is used as evidence in the event of a bad check prosecution, it will be necessary to establish who accepted the check and, in turn, identify the issuer. If you cannot prove by your initials or an employee number written on the check that you accepted the check, prosecution will fail.

Checks to Avoid

A. Counter Checks A counter check is a check which does not have any information printed on it to identify the person issuing the check. There is no account number, name, address, and so on.

B. Starter Checks These are temporary checks given to customers until their printed personal checks are issued. These are brand-new accounts—be careful.

C. Two-Party Checks A "two-party check" is created when a check written to one person is endorsed by that person to a second individual. You would be the second individual in this case. Two-party checks may be good but, if the check is insufficient, collection is very difficult.

✐ Reminder

AVOID—
- ■ Counter Checks
- ■ Starter Checks
- ■ Two-Party Checks
- ■ Postdated Checks
- ■ Altered Checks

D. Postdated Checks If customers ask to give you a postdated check, they are telling you that they do not have the funds in the account to cover your fee. A postdated check is a regular check except that it has been dated for sometime in the future. This check cannot be cashed until that date. Chances are that if the customer does not have the funds now, there will not be funds to cover the amount of the check at a later date either.

E. Altered Checks Do not accept checks that have been altered in any way. If a mistake has been made on the check, have the customer correct and initial it or write you a new check. Do not do it for the customer; that is forgery!

Make Sure It Is a Real Check

A. Check for Perforations All real checks (except U.S. Treasury checks) must be perforated on at least one of the four sides.

B. Check the Federal Reserve Number Is the check issued from a bank or savings and loan located in the correct Federal Reserve District? Check the nine number digit between the brackets along the bottom of the check (Example: 102135783). The first two numbers denote the Federal Reserve District. In Colorado, we are in the 10th Federal Reserve District. For savings and loan N.O.W. accounts the first two numbers should be 30. As a guidepost, the lower numbers represent the eastern portion of the United States and the higher numbers represent the west.

C. Check the Magnetic Routing Numbers The magnetic routing numbers along the bottom of the check must not reflect light. If they do, the check is a forgery or a copy.

D. Color Copies Color-copied checks will easily reflect light and will smear more easily than real checks. The magnetic routing numbers are raised and can be felt like Braille or engraving.

Use Common Sense When Accepting Checks

Ask yourself these questions. Does the identification match? Does the customer seem nervous or defensive about your check-accepting policy?

Look at the check. Is it professionally printed or is it sloppy? Does it contain all the ingredients it needs to be a real check?

If your intuition tells you that something is not right, reject the check; do not accept it.

REMEMBER: You are not required to accept anyone's check, even if you advertise that you accept checks. You may reject any check for any reason.

Collecting on Bad Checks

Now that you know how to accept checks, what do you do if you get a bad check (also known as a bounced check or a rubber check)?

You won't know if you have a bad check until you deposit the check at your bank, your bank sends the check to its bank, its bank returns the check to your bank, and your bank returns the check to you. This awkward explanation is necessary so that you can understand why it sometimes takes at least thirty days before you discover you've received a bad check. At this point, your bank will deduct the amount of the check from your account and add a service charge for processing the bad check. This service charge is normally around $4.

This means that, if the check was for $20 and the bank service charge is $4, you will deduct $24 from your checkbook balance. It also means you did not get paid for that $20 job and, because you were given a bad check, it cost you $4. You will want to collect at least the $24 and try to add your own service charges if you can.

There are three ways to handle this yourself. The first thing you should do is call the customer, explain that the check was returned to you, and try to make arrangements for payment of the bad check, the bank charges, and your returned check fee. This works about 60% of the time.

If this doesn't work you have two options. You can go down to the county courthouse and file a lawsuit with small claims court. This is a lot of hassle, it takes up a lot of time, and there are fees for filing a lawsuit. You will end up spending as much money as you lost and more in time spent trying to collect.

The other option you have is to turn the check over to the District Attorney's office. They will only take the check if you have the proper identification on the check. If the D.A.'s office accepts the identification, they will file charges, take the person to court, and have him pay the court. When they receive payment the court will send you a check for the amount of the bad check. You will still be out any bank charges. This method of collecting is the most effective because criminal charges are filed. If you take the check to court yourself it will be considered a civil matter.

Considering the time, effort, and aggravation involved, we do not recommend trying to do this yourself. There is a better way.

Collection Services

We recommend hiring a check-collecting company. There are quite a few of them out there. You will need to shop around for the best service and price. Here is how they operate. You pay a small annual fee of about $50. Some charge a yearly fee and some only charge a one time start-up fee. When you get a bad check the bank sends the check directly to the collection company and they do all the work for you. They make phone calls, write collection letters, and take court action. They do everything. They make their money by charging the bad check writer a service fee. When they collect you will get the amount of the check and any bank charges you acquired. No lost time. No expenses. This is the way to go and, if you only have to pay a start-up fee, you basically have these people working for you for free!

Some check collection services are also full-service collection agencies and will also handle any accounts you have that are past due. There will be a different fee scale for handling past due accounts.

Keep in mind that none of these methods guarantee that you will be paid. You are going to lose money if you accept checks, but you would lose more money if you did not accept them. This is part of the cost of being the boss!

Out-of-State Checks

Before you accept out-of-state checks you may want to ask your local district attorney if their office prosecutes bad check writers from other states. Many do not because of the costs involved. They may file charges on the bad check writer but will not take any action to pursue it. If this is the case in your area, you will need to decide whether or not you will accept out-of-state checks. We do accept out-of-state checks because we have quite a few tourists during the summer and we would lose too much business if we did not. We have not had problems with accepting out-of-state checks.

Greenbacks/Comchecks

If you ever get a call and the person on the other end of the line asks if you accept greenbacks, he is not talking about greenback dollar bills. A "Greenback" is a check. However, you have to call an 800 number and get an authorization number for it to be approved. These checks, if you get the authorization number, are guaranteed. The 800 number is printed on the check. Comchecks are the same thing, different brand. These checks look funny but they are good funds. Now that you know what they are you can say, "Yes, we accept Greenbacks!"

Money Orders/Traveler's Checks/Cashier's Checks

Traveler's checks are common in major tourist areas. They are always good. You won't get as many money orders and you will probably never receive a cashier's check for your service, but treat all three of these like cash and don't be afraid of giving change. They are as safe as cash, but you should deposit them like checks.

Credit Cards

It will be a great advantage to your business if you accept major credit cards. There are many different kinds of major credit cards—Visa, Master-Card, American Express, Discover, Diners Club, Novus, Bravo, Check

Debit Cards, and so on. There are no differences between these cards as far as accepting them as payment; you will process them all exactly the same way.

The more payment options you have available for your customers, the more customers you will get. Occasionally, the only reason your customer will choose your service over your competition is because you accept the form of payment that they prefer to use and your competition does not. Example: Not everyone accepts American Express or Discover cards. If that's the card your customer uses and you accept it, you have an advantage over the competition.

How Do I Start?

To be able to accept credit cards you will need to have what is called a merchant account. Most people will tell you to look into getting one of these accounts through your bank. This is a mistake. When you go through a bank you are asking them to be the middle man for you by having them transfer funds from the credit card companies on receipts that you have accepted (and that you have already received authorization numbers on) into your account. Banks normally charge a high percentage rate to do this for you and they are much harder to deal with.

Since our goal in this manual is to remove as many obstacles to your success as we can and to make things easier for you through our experiences, we must share a personal experience with you.

When we decided that we wanted to accept credit cards in our business, we were told by the other locksmiths in town to visit with an officer at my bank. We had been doing business with this bank for almost ten years with our personal accounts and about three months with our business account.

We met with an officer of the bank in her office and she began explaining to us that they would require us to write an essay for them that laid out our business plans and included our short-term and long-term goals for our business. She then explained that we would have to provide six business references, a financial statement from an accountant, and our last three years' income tax returns. She actually said these words to us, "If you will jump through some hoops for us . . . ." At that point we ended the interview. We received this same treatment from the next bank and the next, so we decided that we would not accept credit cards at that time.

It's A Fact!

Having a Discover Card can help you get new customers.

Two days later, we received a telemarketing call from Kerri at Discover Card Merchant Services. Kerri asked if we would be interested in accepting credit cards for our business. Reluctantly, we asked what we had to do to qualify for a merchant account. She told us that if we had a Discover Card in our name, we were automatically approved! We did have a Discover Card in our name so that was all there was to it.

If you currently have a Discover Card in your name you are automatically approved for a merchant account through Discover Card Merchant Services. Their phone number is 1-800-347-2000. If you do not have one of these cards, apply for one. This is the absolute easiest way to qualify for a merchant account. There are many other merchant service companies in the yellow pages and you can find them under "Credit Cards." Some of them only require a credit report to open an account for you.

How Much Does It Cost?

You will need to have a terminal in your home or office. Simply enter your transactions and have funds electronically transferred from your customers' charge accounts into your bank account. This terminal works with the phone equipment that you currently own; no additional phone lines are necessary. You can either rent this terminal by the month for around $15 or you may buy it outright for about $250. There are several different types of terminals. Your account representative will help you decide which is best for you.

There is also a fee for each transaction that is based on a percentage of each sale. The average percentage rate is between 1.5% and 3%. Some bank fees are as high as 15%; this is one reason why we don't recommend going to a bank for your merchant account.

These fees are deducted automatically from your merchant account. Example: If your total credit card sales for the month are $1,000 and your percentage rate is 2%, your total transaction fees for the month will be $20 plus (if you are renting your terminal) your terminal rent of $15. Your total cost for accepting credit cards that month will be $35.

At first it may seem expensive that you had to spend $35 just to be able to accept credit cards that month, but you need to consider that $1,000 came from customers who might not have used your service if you did not accept credit cards. Credit cards are safer than checks because the funds are almost guaranteed as long as you call for an authorization number. Therefore, your costs (or losses) are controlled. Compare this with the fact that you never know how many bad checks you may get that you will never be able to collect.

How Does It Work?

Your account representative will train you and answer any questions you may have. Accepting a credit card and entering this information into your terminal is very simple. Customers will normally ask you over the phone if you will accept their card before they ask you to provide your service.

Don't be afraid. Open the car for your customer and fill out your normal invoice and authorization/release forms just like you always do. The only difference this time is that you are also going to fill out a credit card receipt. Always give your normal invoice with the credit card receipt so the customer can turn in your invoice to the insurance company if they are going to be reimbursed for your service.

When you have the credit card receipt filled out, you will call an 800 number to get an authorization number. You will normally reach a recording that prompts you to enter numerical information into your phone. Just follow the instructions and, if the card is good, they will say, "Your authorization number is 123456" and you will write this number in the space provided on your receipt. By giving you an authorization number the credit card company is telling you, "We will pay you." If the card is not good they will say, "This card has been declined by your customer's bank." If this happens you will have to ask your customer for another form of payment. If the card is stolen or if the card has been severely abused, a live operator will come on line and ask you to destroy the card. There is usually a reward of $50 for doing this.

After you have an authorization number, have the customer sign the receipt, give them a copy, and that job is complete.

You have not actually been paid until you enter this credit card information into the terminal at your home and perform a batch upload, which causes the funds to transfer into your account. This money is normally in your account within 24 hours.

 # Charge Accounts

I would gladly pay you Friday for opening my car today! We have heard this many times and we have found it not to be true. We no longer give credit in this way to individuals. Extending credit to your customers will be entirely up to you and at your discretion.

You make all the rules when it comes to this subject. We will tell you the guidelines that we use and why we use them. You are actually loaning your money when you let someone promise to pay you later.

You will need to give credit to some companies and almost all of your larger accounts or you will not get their business. When you are dealing with a national car rental company, for example, they normally are not allowed to make payments from their branch offices. In this case you will probably be getting the requests for service locally and then mailing your invoices to their main office, which could be out-of-state.

You will almost always have to give credit to car dealerships. We are careful not to offend car dealerships by demanding cash payment on our first service call for them because they could become a regular account. It is very inconvenient for them to stop what they are doing and write you a check each time you open a car at their request. Usually they call you because they have a customer waiting to look at the car they are locked out of. The salesperson wants to give all of his attention to the customer instead of taking care of your bill. Most of these places prefer that you send them a statement (statements are covered in the section on **Paperwork**) once a month for everything you've done that month. The best way to handle these accounts is to set them up on a "Due Per Statement" basis.

Per Statement Billing

Per statement billing works this way. Everything is due on the first day of the following month. That means, if you did a job on the first day of the month, that invoice would be due on the first day of the following month. If you did a job on the last day of the month, that invoice would be due the very next day (which would also be the first day of the following month). To use this system you will send a statement that lumps all of the invoices for work you've done for your client in the previous month into one invoice. Do not put your regular customers on a 30-day account. By the time you discover that they are not paying their bills, your invoices will have overlapping 30-day due dates. It will be difficult for you to complain about your bills not being paid if all of them are not overdue. (A common excuse when this happens is, "Oh, I was just waiting for them all to be due—which, of course, will never happen?") When you use the "per statement" basis you will know immediately if your bills are not being paid and you will be able to stop a bad account from getting out of hand.

We do not give second chances to accounts that do not pay our statements on time. If we have a hard time getting our money the first time, we switch them to a "cash only" basis and they will have to pay each invoice as the work is done.

 P.O. Numbers

A P.O. number is short for Purchase Order number. It is normally abbreviated P.O.#.

Many times you will get a phone call and someone will ask if you will accept a P.O. number. You need to realize when you get this call that, if you say no, you could be turning down a very large account that could be a good source of regular monthly income for your business. When a company is asking if you will accept a P.O. number, in most cases, this is the only method they are allowed to use to make purchases. If you say no, they have no other option but to call your competition.

A Purchase Order number (or P.O.#) is basically a record of a purchase. You could look at it like this. Let's say your customer has a tablet with 50 sheets of paper and they number each page from 1 to 50 so that if they ever lose one of the pages, they will know exactly which page they lost. Now, think of yourself as the seller and your customer as the buyer. Here is what happens to the sheets of paper.

Your customer is going to make a purchase from you. Every time they buy something from you, they take one sheet of paper and write the seller's name (that's you) on it. They do this so they will remember who sold them something. They also write down the following information:

- What they purchased
- The date they purchased it
- Where they purchased it
- Who sold it to them
- How much it cost
- Why they bought it, and
- Whether or not someone is supposed to pay them back.

The page number on this piece of paper is the P.O. number.

The need for all this information on one piece of paper is why companies use P.O. numbers. They need to keep track of exactly what they are spending their money for and who is spending the money.

When someone asks you to accept a P.O.#, they are asking you to accept this number instead of immediate payment. In other words, they are asking if they can charge something. They are also asking you to accept a number that will correspond to a written record of the purchase. This is not a guarantee that your bill will be paid but it is proof that the work was ordered. You will write this number directly on your invoice. If the bill is not paid you will have the P.O.# to refer to when you call.

Now that you know what a P.O.# is you can see that one would not be needed for a smaller company. Larger companies that use P.O.#s normally take longer to pay their invoices. It is not uncommon to have to wait 30–120 days to get your money when you have accepted a P.O.#. Once you get into their system you will get a check on time every month.

For More Information

We are not bookkeepers or accountants. We are just trying to give you a basic idea of how the money works. If you have any professional questions that you need to ask, be sure you talk to a qualified bookkeeper, or an accountant, or someone in that profession.

Appearances

SECTION 9

Appearances

Introduction

This section will cover vehicle appearance and uniforms, what the general public expects of a professional, and what you should provide as a professional Car Opening Expert.

If you think of yourself as a professional and dress like a professional, your customers will treat you like a professional. You will get more corporate accounts and have a better relationship with the your customers when your image and your attitude say, "I am here to provide a competent, honest service." Your vehicle advertises who you are, what you do, and what kind of service customers can expect. In fact, most customers will see your vehicle before they see you. Their opinion of your ability may be based on how you maintain your service vehicle.

Be neat, be calm, and be courteous. It will pay off in many ways and when you least expect it.

Vehicles

You should use the personal vehicle that you currently own when you are first starting out. We recommend that you spend as little money as possible on your business start-up. After you see that your business is working, one of the first things you will need to do is to make your service vehicle look like a commercial vehicle. Your customers will be expecting a commercial-type vehicle to arrive. When you drive into a parking lot you want your customer to be able to see you right away. You do not want to have to circle the parking lot trying to find your customer. If you have to hunt for several customers in one day you will use up a lot of time you could have spent opening other cars. It can really put you in a jam if you have other customers waiting.

Some car opening experts use pickup trucks, pickup trucks with campers on them, utility vehicles, or small cars with signs on them. Any type of vehicle that you have will work when you are starting out. If the vehicle you are using has rust spots, dents, missing hub caps or molding, or needs to be painted, you will need to get these things taken care of as quickly as possible. Along with any repairs that you may need to have done, you will also need to have professional signs made for your service vehicle.

The ideal service vehicle for this business is a cargo van. This is a van without windows on the sides. Even the older styles can be fixed up and painted to look as good as new for under $300. This is the type of vehicle that most of your customers will be expecting and watching for when you arrive. The area for signs is large and high for good visibility. The problem you will find with smaller vehicles is that the signs have to be small due to limited space and are so low and hard to see that your customers will have a hard time spotting you.

 Signs

Signs and the appearance of your vehicle are very important. The quality of your service will largely be decided by your customer from the appearance of your vehicle. Sloppy signs and poor vehicle appearance will show that you are not concerned with pride and workmanship. This will also reflect on how your customer thinks you will care for his car. If you are driving a beat-up looking car with hand-painted signs, you will get that customer one time but you will not get his business again. He will call your competition the next time he needs his car opened. Repeat business and referrals are very important to your success.

Don't be cheap when it comes to having your signs made. The cost for having a professional sign made for your service vehicle will range from $50 up to almost any amount you can afford. If you cannot afford to have a professional signmaker make and apply the signs to your vehicle, you are better off without signs until you can afford to have them done right.

There will be times that you will have to allow your customer to ride in your vehicle or just sit inside to stay out of the weather. Sometimes the parking lot is so large that it will be easier to pick them up at the front door of main building and drive them to their car. In this short time your customer is evaluating you for the next time they need your service. Keep the inside of your vehicle clean at all times.

 Uniforms

Your personal appearance is very important. This includes the way you dress, the language that you use, and your personal hygiene. All of these factors will count when your customer forms an opinion of your abilities. You need to look like a professional, dress like a professional, and talk like a professional.

Most of the car opening professionals that we see are wearing jeans and tennis shoes but most of them wear a uniform shirt that has the company name on one pocket and the service person's name on the other pocket. Not everyone can afford uniforms right away so at a minimum wear a nice shirt, clean pants, and do not look sloppy. You will get dirty throughout the day but you should start out fresh every day.

If you are going to wear a hat or baseball cap, have your company name put on it. It is not professional-looking to wear hats with other business slogans or messages on them.

Your language can affect the longevity of your business. If you use bad language when you speak to your customers or if you are continually politically incorrect, they may not use your service again or refer your company to friends and relatives.

Your appearance will not only affect the amount of repeat business you will get, it will also help you on each job. Say you run into a car that is very difficult to open. You are having a hard time. You are a beginner. You have not done very many car openings. If you look sharp, it will go a long way. Your customer will still think you are a professional even though you are just getting started. Why? Because you do not look like a beginner. Looking like a beginner will cause people to have doubts, and you do not want to give your customers any reason to doubt your experience. Look the part. Be the part. You will make more money just because you look good.

Problems

SECTION OUTLINE

Problems

Introduction

Every business has problems and this one is no exception. This business is a very simple one so the problems are not going to be endless as they may seem with more complicated businesses. It makes a world of difference if you know what to expect so that you can deal with them quickly and get on to the fun stuff—making money!

You need to have the number of a tow truck and a mobile mechanic handy to take care of problems with your service vehicle. Avoid opening the wrong car by waiting for the customer to arrive. Except for things like flat tires, crank calls, and what we call no shows (no one is there when you show up), this section covers the major problems associated with this business.

Study this Manual carefully. After you've been in this business for a short time you should have enough experience to handle any problem that arises almost without thinking about it. It is actually going to seem like you have a money-collecting business instead of a service business. Customers will call and you will go to them and open their car so quickly that the work involved is almost incidental. Instead of thinking, "I have to go open a car," you will start thinking, "I get to go collect $35!"

When you first start out it may take you 30 minutes to get some of your cars opened. Most people believe that their car is hard to get into anyway. Don't worry, speed will come with practice and once you get the feel for what you are doing most of your car openings will take about 15 seconds!

We have used bold type and separated these problems so you can find them easily later but you should try to memorize each situation. Remember, your main thrust is to complete each job as quickly as possible so you can get on to the next one.

Reminder

Your goal is to complete each job quickly so you can get on to the next one.

You Cannot Find the Instructions

Occasionally you cannot find the instructions on how to open a car; usually this happens with newer vehicles. Your Car Opening Manual needs to be updated every year to include the cars that have just rolled off the assembly line. Sometimes you just can't find the exact year, make, and model of the car you are working on. With some experience you won't be using the instructions very often because you'll learn to recognize similar features such as: the position of the lock button, the style of the lock button (flip, slide, or lift), how far the lock button is from the top of the door, the body style of the car, the position of the door handle, and so on.

If you cannot find the exact year, make, and model, try another year under that make and model. If this does not work, forget the year and model and look in the instruction book under that make for another vehicle that has a similar setup to the one you're working on. Always look for similarities to help you open an unfamiliar vehicle.

If none of these suggestions help you, try to find another way. This may be the time to try picking the lock open. General Motors products (except for their foreign cars: Geo, Spectrum, Tracker, Storm, Prizm) have locks that cannot be picked open without using a special tool that costs about $235 and is very fragile. We don't recommend purchasing this tool until you have more lock knowledge and experience. If you are working on a car that is not a high-end vehicle or a GM product (Chevrolet, Buick, Pontiac, Oldsmobile, GMC), you will probably be able to pick the lock open. Always lubricate the lock first; this makes picking easier.

 ## You Cannot Open the Car

You are embarrassed because you just cannot open the car. Unfortunately this happens to everyone sometimes but it is rare. Do not be discouraged when this happens to you. You must have a backup plan for this. The best backup is to call another locksmith. Yes, it is embarrassing but you will learn from it. You need to make yourself known to the other locksmiths in town and develop a relationship of working together with them. If you do not wait for your backup to arrive, you have wasted your time and money driving out to that job. Your backup will show you how to open the vehicle and will probably try to comfort you with a story about a car he could not open. Besides, it is rude to leave your customer standing there alone and stranded.

Try to have more than one backup locksmith available (we get busy). Your customer will appreciate the fact that you are taking care of the problem for him so he doesn't have to shop around again. Should none of your backups be available, your only alternative is to apologize to your customer and tell him that you cannot open the car and that he will have to call someone else. This is even more embarrassing and does damage to your reputation. It will happen; you cannot avoid it. It's part of the business. All you can do is to maintain a good attitude and move on. Amen.

 ## You Break the Window

Don't worry about this and don't panic. This is possible but should not be a concern. We have not broken a single car window in all the years we've been opening cars. Always use caution and common sense when dealing with glass. The main cause of window breakage is extreme weather conditions. Glass breaks easily in very cold weather and in very hot weather. You can

break the window inserting your wedges in the window frame if the pressure is too great. Suppose it is very cold outside, but the car glass is hot because the car has been running with the heater on. Your steel tools will be very cold, and as soon as your cold tool touches that hot glass, the glass will pop and shatter. Another common problem in cold weather is removing your tool after you have opened the car. Should the opening tool strike the bottom edge of the glass as you are pulling it out, you will probably shatter the window. Just being aware of these problems will help you avoid them. Always be gentle when opening vehicles and do not try to force anything. If you are having to use force you are not doing something right.

No matter what the weather conditions are or what the circumstances were that caused the breakage, always accept the responsibility for causing the damage. Never tell the customer that it was not your fault. Apologize first, then explain what caused the breakage if you can tell. Complete responsibility is the best policy in this situation.

Sometimes they will be understanding and forgive the damage but do not ask them to. If they do, do not charge for your service. If they don't, do not dismiss your charge for the opening. You are going to have to pay to have the glass replaced and you did open the car. No joke, you did open the car and the customer should pay for your service. Tell the customer that you will take care of the glass immediately by calling a mobile glass service. If the customer prefers, you will schedule whatever time is best for your customer to have it replaced, either at a glass shop or by a mobile service. Allow the option of having your service fee deducted from the repair bill. Again, do not dismiss your service charge. If you have purchased business insurance your policy may cover your losses; ask your insurance agent.

 ## Scratch the Window Tinting

This problem can be avoided altogether by telling the customer before you start that the opening procedure may scratch the window tinting but that this is the only way to open the vehicle. We run into this all the time with after-market tinting. Almost always the customer will tell you to go ahead. Sometimes the customer will say never mind because someone in the family has a key but is not available right now. When this happens we still charge for the service call.

Should you not notice the tinting and accidentally scratch it, always point it out to the customer. Most of the time they will say, "No big deal. Forget about it." When this happens, collect your service fee and go. If you can tell that he is bothered by the damage, offer to not charge for the opening in lieu of the damage. Window tinting is not very expensive to replace (probably $40, tops, per window) so most people will trade the opening for the damage. If not, collect for your service, and ask the customer to replace the tinting and send you the invoice.

NOTE: Always ... Always ... Always leave the customer completely satisfied.

The Lock Does Not Work

You may disconnect the linkage rods in the door if you are not careful and sometimes even when you are careful if you are working in the wrong place. What usually happens is that the clips that connect the linkage rods get knocked off. There is really no damage to the vehicle; it's just a matter of removing the door panel and reconnecting the clips. If the customer tells you the key turns but nothing happens, the lock button works but will not unlock the door, or the handle does not work, this is usually the cause.

Until you learn to do door panel servicing you will have to have someone do this for you. Call your backup locksmith. Stay and watch how he does this and you will learn very quickly. There are only two specialty tools you need to do this and you can get both of them for under $25. Your locksmith supplier also can help you learn this skill; they have books and videos on door panel servicing. It is not hard to learn. The main trick to door panel servicing is learning to recognize where the hidden screws and fasteners are.

False Claims

There are some things you can do to prevent false claims. Your customers will sometimes claim that you caused the damage even if they know you did not. You should always find out if there is a problem before you leave the job. Do this by having the customer check the key in the door while you are there. Be sure he agrees that everything is okay. You should always have the customer sign in two places: once on the invoice where it says, "I acknowledge satisfactory completion of work," and again on the authorization/release form.

Many times the customer will have tried to open the car himself, or one of his neighbors or a passing citizen with a Slim Jim gave it a try. Always ask if anyone has tried to open the car before you. You can tell if someone has tried because the glass under the weather stripping is covered with a film of dirt and will obviously be disturbed and marked up. If you can tell or if he admits that someone has tried to open the car before you, simply say that you cannot be responsible for any damages and have him sign your authorization/release form before you open the car.

We recommend that you only work on the passenger side door. Here are the reasons why. Some of your specialty tools are made to work only on that door. The passenger side is usually away from the street, keeping you out of

> **Reminder**
>
> ALWAYS—Have the customer sign the authorization/release form before you open the car.

harm's way. When people try to open their cars they tend to only try the driver's side door (tow truck drivers do this also). And the most important reason—should your customer call you back with a problem, find out which door has the problem. If it's not the passenger side door you know without a doubt that you did not cause the problem. You can confidently tell the customer your policy is to work only on the passenger side door.

 NOTE: Always check to see if there is a problem before you leave.

 # Nightclubs, Lounges, and Bars

Is this a problem? It can be. The majority of locksmiths that we know refuse to service these places. This is not out of prejudice but from previous bad experiences. Nowhere else will you get the abuse you can receive at one of these places. The experience is the same any time of the night or day. Belligerent customers, and passersby as well, will tell you that you are ripping people off by charging so much even though you have gotten out of bed at 2:00 a.m. to help them back into their car. All they can see is that it took you only a few seconds to get into the car and (after a few drinks) they don't think they should pay for that—and many times they don't. This is a phenomenon that you will not experience anywhere else. These people are not in their right (sober) minds. Our response when they call is "I'm sorry, we do not have anyone available right now."

Should you wish to service these establishments by all means do so. After all, what doesn't kill you makes you stronger. Here's something to think about—You may be liable if your customer is intoxicated and kills someone with his car because, basically, you gave him the car keys.

If you are going to open cars at these places, we do have a dirty trick to put in your bag should you want to get revenge on one of these customers who has refused to pay you for your service. Call 911 from your cellular phone. Report that you are a concerned citizen driving behind what appears to be an intoxicated driver who is swerving all over the road and has almost hit three people already. Give the license plate number and the direction the driver is traveling. You'll be surprised how quickly someone will respond and that DUI ticket will easily burn up $1,500 in attorney's fees. They will send one unit to intercept the driver on the road and another unit to the address indicated on the license plate registration in an attempt to catch the driver as he pulls into the driveway. The customer will never know that you nailed him.

The Customer Refuses to Pay

This is not something that you should be worrying about on your way to every job. It does not happen very often and, when it does happen, it is usually for one of these three reasons:

1. The customer opened the car himself.
2. Another locksmith has beaten you there.
3. The customer did not personally call you so he does not feel responsible.

You can avoid these excuses for not paying your fee by, first, making it your policy to always speak directly to the customer and, second, by being very specific about your fees over the phone. Example: Someone calls and asks, "How much would you charge to come open my car?" If your only response is $25, $35, or whatever you are charging at that time, it is assumed that your fee is for the opening only. Therefore, if you do not open the car, there is no fee.

Develop a standard response such as, "The service call for coming out is (whatever you are charging at that time), but there is no charge for opening the car." What a difference! It is very clear from this response that you expect to be paid for your time spent driving to the location whether you open the car or not. This will save you many wasted service calls and the customer never argues with this as long as there was no misunderstanding about your fee. We allow the customer to cancel, no matter how close we are to the jobsite, as long as we have not arrived and gotten out of our vehicle.

If for some reason the customer refuses to pay, what can you do? Maybe a better question would be, what are you willing to do? A $25 invoice may not be worth the time and trouble of going to court but, if you choose to take it that far, you will win. Choose your battles.

Just going through the motions will sometimes get a customer to pay right on the spot. He is not sure exactly what you are going to do or if he is going to get into trouble. He had assumed that you would just walk away and forget it but, when he sees you doing what we are about to tell you, he usually changes his mind and pays up. Fear of the *Unknown* is very powerful.

The Scenario

First of all, always be very polite. Learn to say please, thank you, Ma'am, and Sir. If you lose your temper and start insulting or cussing at the customer you will lose this money game.

 WARNING: Never ... Never ... Never ... Never, Ever make any threats of any kind concerning physical injury or property damage.

Its's A Fact!

Fear of the Unknown is very powerful!

Explain that the charge was for coming out and that they could have called and canceled your service before you arrived. Then proceed to fill out your invoice and stand in front of the car, bending over to exaggerate the fact that you are writing down the license plate number. At this point they will ask what you are doing. Tell them you have to collect the service call or your boss will take it out of your paycheck (they have no idea how big your company is). Still talking, walk to the back of the vehicle, making it obvious that you are looking for the make and model, and explain that you are going to have to turn them in. Now they are not sure exactly what is going to happen and at this point they usually pay.

If they haven't paid yet, they will ask, "What does that mean?" Remember, you are still bluffing and trying to get them to pay up. Tell them it means that they have received a service and refused to pay. Now you have to turn it over to the police department. The police have to run a license plate check so they can impound the vehicle, place a lien on it, and issue a citation to the owner. Wait for a response. You should be getting your money now. If not then you say, "Oh … and it stays on your credit report for seven years." Most people know about bad credit staying on their report for seven years. This makes everything else you've said sound believable.

 REMEMBER: It does not matter what this person says or does to you, if you walk away with this person's money— You Have Won!

The Result

This bluff has worked for us many times over the years. In reality, all you can legally do is write down the license plate number, make, and model, go down to the department of motor vehicles, and run a license plate search. This will give you the owner's name and address and you will be charged $2 or $3. With this information you can file a claim at small claims court or just turn the invoice over to your collection agency with the search attached. The agency will take the matter to court for you but they usually keep half the money as their fee.

Please do nothing more than this. Let it go after you have filed a claim or turned it over to a collection agency. We know locksmiths who have found the car later and super-glued the locks. This is vindictive. It causes ill will. And it has cost them around $2,500 in attorney fees and restitution plus several hours in court (even though the customer had no proof). They lost credibility, damaged their reputations, and lost time they could have used opening cars and making money!

Getting Business

SECTION OUTLINE

SECTION 11

Getting Business

Introduction

This is the most important section of your Lockout Service Manual and is what makes this Manual so valuable. No other locksmithing course tells you how to build your business in this way. Many years of experience and our observations of other successful locksmiths are presented here. When you are finished with this section of the Manual you will know more about this business than most (if not all) of the experienced locksmiths that you will be competing against!

By using the information here you will soon increase your business so much that you will have too much work. Most of the business building methods here are proactive instead of reactive. Proactive means taking action to make things happen. Reactive means simply reacting to things that are happening to you. Most locksmiths are reactive, not proactive. They depend solely on their yellow pages advertising, passively sit by the phone waiting, and just hope that a customer will call. We have some news for you. The world is a candy store but—if you want the candy—you have to go trick or treatin'!

These are the secrets to success in this business. Apply the suggestions given here. Study and emulate the practices of proven success. You are sure to reap the same rewards!

Learn to Think Like Your Customer

Contrary to what your yellow pages advertising representative will tell you, the yellow pages is not the first place people look for help when they are locked out of their cars. We strongly suggest that you have a good ad in the yellow pages but capturing your customers before they get to that point is what you must try to do. Once your potential customer is into the phone book and has found the Locks/Locksmith or Towing section of the yellow pages, you are competing with every other locksmith or tow truck driver in town. While the customer is in the phone book he will shop around. He will normally call three or four companies and, if they all have the same price, usually the last person called gets the job.

Some of the resources people go through before they will call you are:

1. Call a locksmith or tow truck company they have used before

2. Ask for help from people close by

3. Try to open the car themselves

4. Call family members to see if anyone has an extra key
5. Call the car dealership to see if they have a spare key
6. Call the police for help
7. Call the fire department for help
8. Try to force open the lock with a screwdriver
9. Look for a clothes hanger
10. Ask people close by for recommendations
11. Call roadside assistance or AAA.

Then, if all else fails, they will go through the phone book looking for a tow truck driver or a locksmith to open the car for them.

We want to get a potential customer to call us before they get to the Locks/Locksmith or Towing section of the yellow pages. If we can, this will dramatically increase our chances of getting the job. There are many ways we can do this and one of the more profitable ways is to become a Roadside Service Vendor.

 # Become a Roadside Service Vendor

Most new car dealers now have a Roadside Service program included at no charge with a new car purchase or sometimes a used car purchased from that dealer. This is a free service provided by the dealer that basically covers any minor breakdown on the side of the road such as a flat tire, running out of gas, towing charges, or locksmith services. Sometimes car buyers know this and can call roadside service themselves. Many times they are unaware of the service, and when they call the dealer to see if there is a spare key, the dealer gives them the roadside service number to call. Roadside assistance service is easy for the person locked out of the car. They call an 800 number; an operator asks the nature of the problem and dispatches an appropriate vendor to take care of the problem. The person locked out of the car does not pay for this service. It is absolutely free. The vendor then bills Roadside Service for the fee.

Insurance companies also provide this service for a small fee. With this coverage the insured pays the service provider and is then reimbursed by the insurance company. There are two versions of this insurance. One is called roadside service and the other is called towing. Both of these are pretty much the same as far as your service is concerned and both will reimburse your customer for your fee.

From our experience it seems that the insurance companies are adding this service to all their customers without asking if they want to have it. The fee is small, about $6 for six months coverage, and if you tell them that you don't want this coverage they will argue that you need it! Should you complain that you have been paying for this service and did not want it, they will say they were doing you a favor! We believe insurance companies are making a lot of money by insuring many people without their knowledge, while people are paying for these roadside problems out-of-pocket because they do not realize they have insurance coverage. If you will tell each potential customer who calls that their auto insurance may cover your fee, this will increase the number of jobs you get. Be sure to tell them that if they have this coverage there is no deductible to pay and that it does not affect their premium. The insured is normally allowed to use this service twice a year at no charge.

Roadside Service accounts can be very large. Just one account could be worth $50,000 a year or more depending on the population in your area. All you have to do is call the roadside service company and ask them to send you an application to become a vendor. Each service may have different guidelines that you must agree to, such as being a 24-hour service and available 365 days a year, a member of the Better Business Bureau, able to meet their insurance requirements, and so on. Try to comply with whatever they ask of you; these accounts are worth any amount of jumping through hoops in order to land one.

Here is a list of roadside service providers for you to contact for applications. Some of these companies may be grouped together into the same service contracts; you will learn more about their policies by calling each number. If these numbers change, find the dealers in the phone book, call them, and ask for their roadside service number. (**Believe it or not, many locksmiths are unaware of these contacts.**)

AAA	1-800-283-5AAA	Jaguar	1-800-452-4827
Acura	1-800-862-2872	Jeep	1-800-533-7324
Alfa Romeo	1-800-245-2532	Lexus	1-800-872-5398
Audi	1-800-367-2834	Lotus	1-800-245-6887
BMW	1-800-334-4269	Mazda	1-800-639-1000
Buick	1-800-422-8425	Mercedes	1-800-622-7550
Cadillac	1-800-333-4223	Mitsubishi	1-800-521-4140
Chevrolet	1-800-243-8872	Nissan	1-800-647-7266
Chevy Truck	1-800-962-2868	Oldsmobile	1-800-242-6537
Chrysler	1-800-422-4797	Plymouth	1-800-759-6688
Corvette	1-800-222-1020	Pontiac	1-800-762-4900
Dodge Truck	1-800-423-6343	Porsche	1-800-252-4444
Eagle	1-800-533-7324	Saab	1-800-582-7222
Ferrari	1-800-447-4700	Saturn	1-800-522-5000
Geo	1-800-243-8872	Subaru	1-800-782-2783
GMC Truck	1-800-462-8782	Suzuki	1-800-447-4700
GM of Canada	1-800-263-3777	Toyota	1-800-468-6968
Infiniti	1-800-826-6500	Volkswagen	1-800-444-8987
Isuzu	1-800-792-3800	Volvo	1-800-458-1552

 Tow Truck Drivers

The next most profitable way to find customers is with referrals from your competitors! You will have obvious competitors but keep an open mind in this area. Anyone who opens cars is your competitor whether he is charging for the openings or just being helpful.

A tow truck driver might not be someone you consider being in your trade but they do open cars and they open a lot of them. They can be a valuable asset to your business. Unfortunately, tow truck drivers and locksmiths have had a bad relationship for some time. It seems that in general, locksmiths do not feel that tow truck drivers should be allowed to open cars at all, simply because they have not been formally trained in car opening methods. In truth, the locksmiths have had no formal training either! Some locksmith trade magazines advertise T-shirts for sale with slogans like, "If you want your car opened call a locksmith. If you want it ruined call a tow truck driver." There are other slogans on T-shirts that insult the towing industry as a whole. We don't want to list them all because we are not promoting these prejudices. If this is the attitude that the locksmiths in your area have toward tow truck drivers then establishing a good relationship is an untapped gold mine for you!

In general, tow trucks are called for towing. The number of calls they are getting to open cars is not as great as the number of towing calls. Therefore, many tow truck drivers do not understand the importance of owning the best car opening tools they can get. They don't understand that the cost of a good set of tools can be earned very quickly. We have been on many car openings where a tow truck driver has been there first. The reason for this is because most of the time they just do not have the right tools and have to tell their customer to call someone else. We are sure that they would rather not just tell the customer to call someone else. This is rude. However, they will refer the customer to your company if you can develop a relationship with them. We have some towing companies that are referring lockouts to us on a daily basis! One of these towing company owners has told us that he understands the need to have the right tools and, since he owns five tow trucks, he would rather not spend $1,200 to outfit all five vehicles. **He prefers to send all of his lockout customers to us!** We do not understand this reasoning but hey, it works for us!

Because of past bad relationships between locksmiths and tow truck drivers you will need to approach these people carefully. We suggest you approach them in the following manner. It will be well worth your while.

Do not call them on the phone; this will get you nowhere. You must visit with them in person. As soon as you introduce yourself as a locksmith they are going to have their guard up. You will have to bring that guard down before they will hear anything you have to say. This method works very well for us. Walk into the office. The very first thing you are going to say is, "You guys have all the fun." Laugh a little and then add, "I've always

☞ It's A Fact!

It is hard not to like someone who likes you.

wanted to be a tow truck driver, I just don't think I could repossess someone's car!"

It is hard not to like someone who likes you. Right? Make them feel important—you can do this by being interested in their trade. Now introduce yourself and your company and say, "I know you open a lot of cars and I would like to be your backup for lockouts when you get too busy." Leave some business cards and you're done. Do not claim to be better at opening cars than they are; this will only fire up those old feelings. Do not claim to have better tools or tell them they do not have the right tools. Do not claim to be better at anything that they do, period. It sounds silly, but you want them to accept you as an honorary tow truck driver. (Ask them to do this; they will get a kick out of it.) If you can do this it will separate you from the other locksmiths and that old rivalry. You have become an associate, maybe even a friend.

 ## Locksmiths

Your two best sources for referrals from your competitors are going to be tow truck drivers and other locksmiths because these are the two main sources of car opening services available to the public. You will actually get more referrals from the tow truck companies because there are usually more of them in town.

You should stop by each locksmith shop in your town and introduce yourself.

 SMOOTH MOVE TIP: It helps to come bearing gifts such as donuts, cookies, or other goodies. Also call the locksmiths that are "mobile only" and ask to meet with them for lunch or coffee.

It is very important when you are talking to a locksmith never to claim to be the best in the car opening business or to have better tools. You need to remember that you have fallen into the "gravy" part of this business and that most locksmiths have had to go through extensive training to be able to run their businesses. Some locksmiths may resent the fact that you only get the "gravy" jobs while they have to deal with the more tedious work associated with locksmithing. You may be making more money than they are; don't rub it in. It will help your relationship to let them brag. All you need to do is show an interest in what they are saying and doing. For the sake of your relationship, when they ask you how your business is doing, no matter how much money you are making, always say, "Things have been really slow."

Reminder

Always let the other guy have bragging rights. It works just like catching more flies with honey than vinegar.

Try not to give the impression that you are only interested in what they can do for you; this will shut them out completely. You want to develop a relationship with them because both of you could profit from working together.

Start by explaining that yours is a new business and that you are a one-man service. Next, ask them if they would be interested in taking any jobs that you could not get to if you are too busy. You need to develop relationships with several locksmiths so that you will have a backup system in place should you ever need help on a job or need to refer one of your customers to someone else because you get too busy.

You will get many referrals from the locksmiths in town just because most of them are on call 24 hours a day. Everybody needs some time off. They will refer customers to you at the end of the day and on weekends when they are tired or burned out.

Police Officers

We are astounded at the number of people who think they can call the police department to open their cars for them. We are even more astounded that the police will sometimes actually do it!

Call your local police department and ask them if they help people who are locked out of their cars. Not all police departments have the same policies and policies change all the time. If you have more than one police precinct in your town, make sure you call them all.

If your police department does not open cars, they get so many calls for this that sometimes they will have a vendor list. They will actually dispatch a vendor from that list for the person locked out of their car or will give a phone number to the caller. The person locked out will pay the vendor directly for the service fee. If they do have a vendor list, you want to be on it! Ask them to send you an application to get on the list. You will probably have to go down to the station to pick this up. If you can get on one of these lists you will be "in" with the police department and you will get many jobs from this source.

Some police departments have a policy that no police officer on the force is allowed to open cars for the sole purpose of helping someone back into his car (to prevent damage liability), but they are allowed to open cars in emergencies. A child or animal locked in the car would be considered an emergency. If they do have a "no car opening" policy, it means that they will not dispatch an officer to open your car for you. Nevertheless, police officers are just like anyone else and if you can flag them down on the street they will normally try to give you a helping hand. Almost all of them carry a car opening tool called a Slim Jim for emergency car opening situations. This is the only car opening tool they carry and this tool is extremely limited in what it can open. Because of this, police officers are not successful very often when they try to open a vehicle and must recommend that you call someone in the car opening business for help.

Police officers find themselves in this situation all the time because people expect car opening assistance from the police. Many people believe that the police are trained in car opening and in breaking and entering techniques. You might think that they would have to know these things to investigate some crimes but this is not true.

There is a lot of power associated with being a police officer and you must approach them properly when trying to get referrals from them. They may have a policy that they are not allowed to recommend a specific vendor because some people in the community would see this as being unfair to the other vendors. No matter what their policies or rules are, we are telling you that they will open cars and they will recommend specific vendors whenever they feel like it! Remember these are very powerful people.

Flattery goes a long way with these people. This sounds bad but it is true of all of us. What we mean is: the police spend most of their time dealing with bad guys and trying to help people who sometimes don't want their help. The abuse that they take on a daily basis is incredible. We don't think they hear anyone say "Thank you" very often. Any respect that you can show them will be appreciated.

When you approach them you need to give the impression that you are trying to be of service to them instead of trying to get something for yourself. Of course, you are trying to get business from them and a lot of it. This will happen if you approach them correctly. Do not ask them to tell people to call you when they are locked out their cars. If you do, we guarantee you—they never will.

Start your conversation off by asking if they ever have had to open a car because a child or animal was locked inside. Then introduce yourself and your company and tell them (as you hand them your card) that, if they ever have trouble getting into one of these cars, you will provide this as a community service at no charge. You want them to know that you are available as a backup system for them.

Another approach could be asking to be their personal locksmith. We've found this phrase to be almost magical with anyone you are trying to get business from. Use this phrase as often as you can when offering your service. Actually say to the person, "I would like to be your personal locksmith." This is very powerful. If you say, "I want to be your locksmith," without using the word personal it sounds money oriented. When you add the word personal, it is more like you're saying, "I care about you." This is very important.

 REMEMBER: People do not care what you know, until they know that you care.

Tell the police officers that you would like to be their personal locksmith just in case they ever find that they personally are in need of lockout service. Make your offer or introduction on a personal level—just you and the officer, no one else. It is important when you are recruiting police officers to refer business to you that you never ask for anything, not even advice on any questions concerning the law or your rights. **Ask for nothing, absolutely nothing.** If you use this method you will start getting referrals from them. You might get many referrals from a single police officer in the course of a week.

You will need to introduce yourself to as many police officers as possible. How do you do this? How many donut and coffee shops are in your town? The police and donuts have been a joke for a long time but it is true that they spend a lot of time at these places. Start hanging out at these places when you have nothing else to do. Find out when their scheduled breaks are and make it a point to be at the coffee or donut shops at that time.

Fire Departments

Fire departments also get calls from people locked out of their cars but will only respond if a child or animal is locked in the vehicle. We really do not understand why they respond at all instead of referring the caller to a car opening service, unless they feel that the child may need medical attention. Even then they seldom call an ambulance to assist them on these calls. We have been on many car openings while the firemen are there and they normally continue working on the car until you actually open the door. By that time they are working on the car just for the fun of it. They enjoy doing this as much as most people do but they usually have only one car opening tool, the Slim Jim. Just to remind you, the Slim Jim is practically useless on today's newer vehicles and its use is extremely limited.

Firemen are not asked to open cars as frequently as police officers because they are not cruising the streets and therefore not as visible. However, many people expect a fireman to be able to open their cars and they do get many calls requesting free car opening service. Firemen are much easier to approach and recruit for referrals. You do not have to be as careful of offending them but use some of the same strategies that you use with police officers.

Since you don't see many of these guys on the street you will have to visit with them at the fire station. All the firemen that we have met have been very friendly and helpful. Just walk into the station (there may be many stations in your town) and begin introducing yourself to the people there. Tell them that you know that they get calls from people who have locked an animal or child in the car and that you want to make them aware of your service. Tell them that you do not charge for your service in this situation. Do not be afraid of offering your service for free to the community in emergency situations. You may have to open some cars for free, but you will get many paying jobs by doing this. The fire department gets car opening calls that are not emergencies and will refer your company if you develop a relationship with them.

Most locksmiths don't think that police officers or firemen should be allowed to open cars at all and will not work with them by sharing information or advice on tools. We've found that police officers seem to be perfectly happy with their Slim Jims but that firemen are not.

You can begin your relationship with firemen by offering to update their car opening tools (at a profit). Ask them what they have for opening tools and ask if they would be interested in getting a better set. You will find that most of these guys are very interested in car opening. If you let them know that you are willing not only to sell them the tools but also to show them how to use them, you will develop a relationship that will be very profitable for you. They are only allowed to respond to emergencies and you will get the rest. They may have a policy that they are not allowed to recommend a specific vendor, but if you work on your relationship you will get referrals from them.

Car Rental Companies

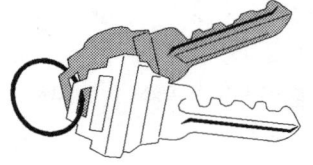

You may be surprised to learn that most car rental companies do not keep a spare set of keys for each vehicle. Usually the renter drives away with the only set of keys to the car. This is common policy with most car rental companies.

The facts are that the renter is not familiar with the vehicle (most of the cars are brand new) and that having to rent a vehicle is a distraction from their normal routine. They have other things on their minds (their own car has broken down or they are on vacation) and are not paying attention to the rented vehicle. This stress causes many people to accidentally lock themselves out of their rented cars. Also, because of carjackings, most of the newer cars are built to lock all the car doors automatically when you start the engine. Lots of people are unaware of this new feature so many are locked out when they leave the engine running to go into a convenience store, restroom, or gas station. The very first thing that the renter will do is to call the car rental company and ask if they can send someone out with a spare key to open the car for them.

Car rental companies have a vendor list for this purpose. When a customer is locked out of the car, the rental company will call all the vendors on the list until they find someone who can respond quickly. The renter usually pays the service fee but sometimes you will bill the rental company. Always ask who pays when you get a call from a car rental company.

All you have to do to get on one of these vendor lists is to call each car rental company in your area and ask for an application to be a vendor. Be sure that you call each office and each location. You may have two or more car rental companies with the same name that could be individually owned franchises. Don't forget the ones in the airport terminal if you have one in your town. We occasionally get calls from out-of-state car rental companies to assist one of their customers who locked themselves out while visiting our town.

Hotels/Motels

Many of the guests at hotels and motels are driving rental vehicles and most travelers do not stop to check into a room until they are too tired to drive any longer. These two factors result in many lockouts. This is a good place to get business. The first place a person goes for help when he is locked out of the car and staying at a motel is straight to the office. The first question he asks the desk clerk is, "Do you know anyone who can open my car?"

In the smaller hotels/motels you'll want to stop in and talk to the desk clerks. Introduce yourself and your company, give them your card, and ask if you can place your card on the bulletin board and in the pay phone area. It will help if you have some larger than normal cards made for this purpose. There are many places you can use these larger cards. Be creative. Some locksmiths choose to have flyers made for this purpose but most hotel/motel managers will remove them as soon as you leave. They feel that flyers are intrusive and clutter up the place. Have some cards made that are about the same size as an index card and keep your advertising simple.

 DETOUR: There are only two times it not wise to brag about your car opening skills or your company when you are trying to get referrals. You already know when those times are: talking to tow truck drivers and talking to other locksmiths. This is sure to build rivalry between you, and they will not refer any customers to you. With everyone else it will help build confidence in your company if you tell them that you are the best in the car opening business. Tell everyone but tow truck drivers and locksmiths that you and your company are the very best, that you can open any car on the road quickly, that you guarantee no damage, that you can give a quick response time, and that you have very competitive prices.

The larger hotels/motels are more service oriented than the smaller ones and are a great place to "lock in" your customers. Most of the time when a guest is locked out of his car the Hotel Gigantic will take care of the problem for him. The person locked out will not even have to use the phone. Many times a guest will call the front desk and say, "I'm locked out of my car. Will you call someone for me?" No sooner said than done.

Many of the people working at these hotels/motels rely heavily on tips for extra income and taking care of problems like this is a good source for tips. You will need to make yourself known at the front desk and leave your card. You will also need to meet each bellboy, doorman, hostess, and server, and visit with every person on each different shift. Ask the front desk when the shift changes are and come back on each shift to introduce yourself to everyone all over again. Don't forget, someone else will be at the front desk. Hand out those business cards and leave extras for the personnel at the front desk.

If you can respond quickly and perform your service in a professional manner you will make these people look good, they will get more tips, and you will get more referrals.

Bulletin Boards

There are many free bulletin boards all over town where you can post a flyer or leave an index-sized card. You should use all of them. The best ones are at the supermarkets and larger superstores because they have lots of cars in the parking lots. Do not just post your stuff on the bulletin boards and leave. When supermarket and superstore customers get locked out they ask the store manager for help (or ask if they can borrow a clothes hanger). Always go into places and ask to speak to the manager. Everyone knows that these boards are open to the public but ask the manager for "permission" to post your advertisements. This is a good way to get to meet the managers and make them aware of your services. The next time they are asked to help locked out customers, they can helpfully refer them to the bulletin board instead of turning them away.

Security Guards

Security guards (for the most part) are seen by the public as police officers and they are asked to open cars just as often. Your best source for referrals from security guards is at the shopping malls. The security guards at the malls get so many requests to open cars that they have usually purchased a set of opening tools in order to be of service to the mall customers. We have never seen one of these guards with a good set of tools and their success rate is not very high. Shopping malls usually have a security office where the security supervisor stays, most of the time, to dispatch the other guards over a walkie-talkie system. The other guards are normally circling the mall performing traffic control, giving directions, and so on.

Start at the security office by handing out cards and telling everyone about your service. Try to talk to each guard and give each one your card. Do this during the day. If you are in the area late at night, drive around the mall and find the guards in the parking lot. This will be another work shift and you'll need to introduce yourself again.

Many businesses hire security guards to drive by and check their stores several times during the night. Look these security companies up in the phone book and stop by their offices to hand out your cards. You will soon begin getting referrals.

Car Dealers

Used car dealers will be glad to meet you and welcome you to their lots. They get locked out of their cars on almost a daily basis. They usually do not have spare keys and, between their salesmen and the customers driving the cars several times a day, they are sure to get locked out. Most of these lots already have vendors they use on a regular basis. You'll need to ask for the opportunity to be their backup just in case their regular vendor cannot get to them promptly.

Being locked out is usually an emergency situation in the eyes of the dealer. He is afraid that, if someone wants to look at one of the cars and it is locked, he will not be able to show it and will lose a sale. The dealer's main concern is how quickly you can respond when called. The very first time their regular vendor cannot show up on time you will have a new account.

This is such a good source for car opening business that almost everyone in our business visits these places trying to capture their accounts. You will have to be more competitive with your prices for car dealers. It won't take very many of these accounts to keep you busy.

New car dealers do have spare keys to their vehicles so only occasionally will you get a job or two there. Do visit the new car dealers; most of them have used car lots. If the salesmen at the new lots are impressed with you, they will often call the used car lot and recommend that they begin using your service.

Car lot accounts will come and go and will go back and forth between vendors. You will need to develop a close relationship with the salesmen at these places. These are the people who will be calling you.

 KEEP THIS IN MIND: Most locksmiths are only interested in developing a relationship with the managers. They think that this is where the business is. Wrong! It is the "little guys" who will be calling you.

Stop by whenever you are in the area and ask if there is anything you can do for them. This will help build your relationship.

Parking Lots

Anywhere that people are parking their cars is a good area to post your advertising or hand out your cards if there is someone close by. When someone gets locked out of his car he normally does not head straight for a phone to call for help. He will look for someone close by to ask for assistance. This is the trick to capturing your customers before they reach the yellow pages. Begin by looking for the larger parking lots that have at least 150 parking spaces. Some of these places will be department stores, shopping malls, hospitals, sports facilities, colleges, supermarkets, and large employers. Stand in the middle of the parking lot and look around. You are looking for places people will go to ask for help. If the closest place that you can see is a convenience store six blocks away, that is what you are looking for. There will normally be more than one place to go to where you could ask for help. Visit each of these places, hand out your cards, and ask if you can post an advertisement near the phone. At the smaller places you will want to give a card to each employee if you can. Don't forget to tell each person that you would like to be "your personal locksmith."

Express Service Area

Purchase a fold-out map of your town and circle the area around your "home base" or the area that you will be in most of the time when you are not out on a job. This area needs to expand out from your "home base" to an area you can respond to in fifteen minutes or less. This will be your "express service area" and this is the area you will want to saturate first.

Tell each potential customer or referral source in this area that they are in your "express service area" and you can respond very quickly to them. This will be a good reason for them to use your service. However, the main reason you are saturating this area first is to get more business in less time. You are not limited to an hourly wage rate of income in this business. The more cars you can open in an hour, the more you will increase your hourly rate of income.

If you can open three cars in one hour and charge $45 an opening, you've just made $135 an hour! This does happen—not all the time—but it does happen! We don't know why, but it seems that most of the time you don't get just one call at a time. Calls seem to come in groups of two or three.

Telemarketing

Telemarketing works very well in this business but it is not getting on the phone and blindly dialing numbers. The best way to use this method is set up in five phases. You are going to target your market and you will get many

regular accounts from doing this. First, choose a target. Example: Hotels, or used car dealers, or car washes—any place that can use your service on a regular basis. Choose only one target at a time. These are the five steps you will need to go through and you must do them in this order:

1. Choose a target market.
2. Design a flyer that speaks specifically and directly to that market.
3. Personally visit each location and introduce yourself and your company.
4. Mail your flyer.
5. Make the phone call.

It is important that the phone call be made within two days of the mailing. If you wait longer than that you will lose the impact of this combination punch. The phone call is simple. Just say, "Hello, this is (your name car opening service). Just wanted to know if you received our flyer and if there is anything that we can do for you today?" This kind of approach is starting to make a comeback and people love it!

Your goal is to get people to remember your name. You will probably have to repeat this mailing at least seven times before they will remember your name. This is another trick to getting customers. Don't blindly send mail to everyone in your town one time; everyone will forget you very quickly. Choose your customers. Keep a list of the customers you would like to have and keep hammering them! Mail your flyers to the same customers over and over again, about every two weeks or so, until you begin getting business from them. Always follow up with the phone call no later than two days after the mailing. This is very important—this tells the customer you care.

 # A Word of Caution

If you are not yet listed in the yellow pages or are starting your business before your yellow pages ad is distributed, you will have to use these methods of getting business in order to be successful. When you are in the yellow pages these strategies will dramatically increase your income. If you have other ideas that we have not talked about here, feel free to try them but be very careful when it comes to spending your money on advertising. We have tried or we have seen our previous employers (people we have worked very closely with) try every possible advertising trick in the book. The only methods that work effectively (besides the yellow pages) are the ones we have explained to you in this section.

We would just like to caution you again that we have tried and, in some cases, relied very heavily on the following sources to generate income. None of them even covered the cost of the advertising. These sources should be considered only after you have become successful. Their main benefit to your business is to develop name recognition. Name recognition is very important to your business and is developed over time. These sources are good for developing name recognition but are not good for immediate measurable increases in your business.

Television	Radio
Newspaper ads	Blind mailings
Booths at trade shows	Presentations to special groups

The only things that work immediately in this business are the yellow pages advertising and the "footwork" that we have laid out for you here. We believe that this is due to the fact that this is an emergency service business and people do not plan on ever using your service. They don't think it will ever happen to them, until it does.

Many have failed in this business simply because they concentrated their efforts in the wrong places. Television, radio, and blind mailings are expensive but do have their place. We have covered the main areas where people will go to look for help when they need you. This is where you need to concentrate your efforts. If you work hard to cover these areas you may find that you can eventually reduce the size of your yellow pages advertising. If you can get people to keep your business card, they will call you when they need you.

You must become a professional business card dispensing machine. Tell everyone you meet that you are in the car opening business!

Telephone Skills

SECTION OUTLINE

SECTION 12

Telephone Skills

Introduction

You and your telephone are about to become inseparable. You are joined at the hip. Your telephone has become your best friend, your worst nightmare, and your source of income. How you conduct yourself on the telephone influences your customer. Try to always remember that you are talking to a stressed and distressed person. Your customer is embarrassed, inconvenienced, and nervous in the best of circumstances. He may also be cold/hot, angry/frightened, late/very late, hungry, sick, or in trouble at home. You are a rescuer, a hero, and everything in life (at that moment) depends on you being able to save the day. Impressive! Now let's talk business.

Answering the Phone

Remember, you are a mobile service and you can appear to be a large corporation, a small mom-and-pop shop, or a fly-by-night con operation. It all depends on the level of professionalism you use over the telephone.

A very large part of your appearance is your appearance on the telephone. It sounds kind of funny saying "appearance on the telephone," but how you appear to your customers, how they visualize your company when you are on the telephone, is a very important part of this business.

Believe it or not, we've heard locksmiths answer "Yo!" These were not professional locksmiths! If you simply answer "Hello," the caller begins asking himself these questions: "Do I have the right phone number? Am I talking to the right person? Did I dial the wrong number? Should I hang up and try again to see if I get the right company?" He may just hang up on you and go through the phone book looking for a different car opening company. If you answer professionally you will get a lot more business. Answer in a professional manner such as, "Hello, (announce your car opening service name)." Then say, "This is (your name); how may I help you?" or, "How can I help you today?" If you only answer, "Hello, how can I help you?" you do not appear as professional. There is quite a bit of difference between these two answers. One gives your caller the impression that you've answered another call, you don't want to do it, you're tired, it's late, and you just want to go home. The other sounds like you are saying, "I would love to help you–I'm ready to go!"

You need to have enthusiasm (pep) in your voice when you answer the phone. Do you remember the saying "People don't care what you know until they know that you care"? Practice answering in the same tone of voice every time. Be enthusiastic, caring, and ready to go! I once worked for a locksmith who was very good at this and while under his employment I was required to ask customers if they had called around for prices and why they chose our company. You would be surprised at the number of people that told me "No, you did not have the best price but the man on the phone wanted our business more than anyone else!" Price is not everything. There is more to this business than having the lowest prices.

Reassuring the Customer

All of your callers have just locked their keys in the car. They feel stupid. They have to call someone they don't know to ask for help and you need to make them feel comfortable. Never say anything derogatory such as, "Well, stupidity is expensive" or "There are a lot of dumb people in the world." If they tell you, and most of them do, that they feel so stupid for locking the keys in the car, you need to defuse those feelings immediately. If they are that embarrassed and you can make them feel comfortable they will be less likely to call someone else. We usually do this by saying, "Oh, I do that all the time" or "Everybody has to do that at least once." It is not worth losing a repeat customer by offering someone who has used your service before "The Double Dumb-Ass Discount." There are so many rude people in this business that if you act professionally the market is practically wide open.

Getting the Information

We have noticed over the years that, the longer you can keep a customer talking to you over the phone, the more likely it is that you will get the job. We haven't figured out why. Maybe he spent so much time telling you everything that he doesn't want to go through that all over again with someone else. Or maybe the time spent with you on the phone has created a sense of trust. We know it is true. Ask what kind of car it is, what color, what part of town the car is in, if the keys are actually in the car, if the car is running, and so on.

He is on the phone and locked out of the car. He may not have made the decision to use a locksmith yet but is checking to see what options are available. Tell him he's probably covered by insurance and tell him you are fast—he may decide on you.

Insurance

Tell them your service is reimbursed by most insurance companies. If they have full coverage insurance they most likely have either roadside assistance coverage or towing insurance or both. Either of these will reimburse your customer for your fee. There is no deductible for roadside assistance or towing coverage and it will not affect the premium. Make sure you tell them this. You will probably be the only person that has told them, they will be thankful for the information, and you will get the job. The customer pays you, turns in your receipt to the insurance company, and the insurance company mails them a check for the full amount. Most insurance companies allow them to do this twice a year at no cost.

Response Time

You will soon discover that you are in an emergency service business. Why is a car opening an emergency? It is not really, but if you do not treat it this way you will lose a lot of customers. In the town that we work in, if we have to give a response time longer than 15–20 minutes, 8 out of 10 times we will lose that job. Our customers simply do not want to wait longer than that. But that does not mean that they won't.

It will take you some time to perfect your telephone skills so that you can adjust to what your customers are telling you. Listen carefully, some want to be assured that you know how to open their cars while others are more concerned with how fast you can get there. There are those who only care about price. If you are listening carefully and can tell what it is that they are most interested in, you will get more jobs and be able to charge more. An example of this is: We answer the phone and the **very first** words they say are "how long" or "how fast." We know we can charge more by giving a short response time. On the other hand if their **first words** are "how much" or "what do" we know they are shopping for price and we are going to have to be competitive. Working the telephone is a skill and will take some time to master.

The response time in your area may need to be longer or shorter than 15–20 minutes. You can get a good feel for what is acceptable in your area by calling the local locksmiths and asking how soon they could get someone out to open a car for you. Many times we know when we tell our customers we'll have someone there in 15–20 minutes that it is going to take us 30 minutes to an hour to get to them. You must be willing to do this if you are going to make money in this business. The reason this happens is because it seems that most of the time you will get two or three calls back to back. If you have call waiting, you will still be on the line with the first customer when you get another call and then another. These three calls may be all the calls you are going to get for the next three hours so you need to do what we call stack 'em, rack 'em, and knock 'em down. This means telling all three customers that you will be there in 15 or 20 minutes and then doing them all as fast as you can. This can be very stressful and happens regularly. It is the nature of this business. We realize that this may seem ugly to you, but this is a business and these are the facts of life.

Almost every customer who calls will, sooner or later, ask you how long it will take you to get there. It is very useful to have a stop watch and time yourself from the time you hang up the phone to the time you reach the customer. When the customer calls and says it has been an hour since they called, you can give them the actual time. It seems to make them relax and give you more time. The customer always exaggerates how late you are.

Very rarely does a customer get upset when you are late, but they will call you in exactly 15 minutes and say, "I called you almost a hour ago and no one has shown up yet. Are you on the way?" You should come up with a standard apology and say that the service man (they don't know that you are the service man) should be there any minute now. They will wait if they think you are just around the corner because they don't think anyone else could get there sooner. If a customer calls and is really angry because you are late, they will always calm down very quickly if you apologize and offer to give them a discount because they had to wait (usually customers are very happy to get $5 off). We only give this kind of discount if they are really angry and this is very rare. Most will be slightly irritated that they had to call you again but they are all happy to see you when you arrive.

The Law

SECTION OUTLINE

SECTION 13

The Law

Introduction

It is your responsibility to learn the law. This means all the laws that pertain to your trade, the way you do business, how you order supplies, and your liability. Learn local, county, state, and federal laws. Just knowing where you stand will protect you and your customers and will give you peace of mind. If you are on your way to establishing a good relationship with local and state law enforcement officers, you will find them a good source of information.

Nevertheless, learn the law. Don't take anyone else's word unless you are positive the information is accurate and current. Laws change; be sure to keep up on the latest information and protect your name and reputation. One of the best ways to protect yourself after you learn the law is to keep complete, accurate records.

 ## Ignorance Is No Excuse

Both the court system and the IRS have adopted the policy of "Ignorance is no excuse!" and you will hear this almost every time you have to deal with them. "Ignorance is no excuse" means that you will not be given any consideration or leniency in your sentence, penalties, or fines just because you were not aware that you were breaking a particular law or rule. As far as the law is concerned, as a locksmith you are expected to protect the safety of the public and to exercise due care in the operation of your business. You are also expected to know what you may do and what you may not do in the interest of protecting the public as you provide your service.

 ## What Is Due Care?

Sounds like "Do Care," doesn't it? Basically, as far as you are concerned in operating your business, this means that the law requires that you do care about the safety of the public you are serving. But just caring is not enough. You must show proof that you care. You must prove that you are only opening a car for the owner of the vehicle or other authorized person by obtaining proper identification of the person requesting your service. And you must keep a record of this in writing. Your Authorization/Release form will do this for you (covered in detail in the Paperwork section).

If you cannot get adequate proper identification you must use other means such as having someone else vouch for that person (and getting their identification), asking verification of identity from neighbors, the landlord, an employer, and so on. Always write this information on your invoice or authorization/release form.

 ## Can You Be Criminally Charged?

If you were to open a car for a thief or other unauthorized person, could you be charged as an accessory to the crime? Unless the court is able to prove that you were negligent in the conduct of your business or had conspired with the thief or unauthorized person, no criminal charges could be filed against you. A locksmith who requires all customers to show a license and registration or asks other persons to identify the customer before the car is opened is, for all practical purposes, exercising "Due Care" and could never be charged with negligence in the conduct of his business.

 ## Postal Laws

According to the United States Postal Service, all locksmith devices mailed in the United States are not allowed to be mailed through the Postal Service unless addressed to any of the following:

- Bona Fide Locksmith
- Bona Fide Repossessor
- Lock Manufacturer or Dealer
- Motor Vehicle Manufacturer or Dealer.

If you do not fall within one of the categories listed above, your supplies must be shipped via United Parcel Service (UPS) or Federal Express. Unfortunately, UPS is more expensive and takes longer to deliver than the United States Postal Service.

 ## Possession Laws

It is the responsibility of the owner of potential "burglar tools" to ascertain and obey all applicable local, state, and Federal laws in regard to possession and use of these tools.

It may be against Local, City, Township, County, State, Federal, or Other Laws to Own, Use, Carry, Conceal, Purchase, or Have in your possession some or all of the tools necessary to perform your lockout service.

These include:

- Lock picks
- Car opening tools
- Crowbars
- Hammers
- Glass cutting equipment or
- Any other tool used to commit a burglary.

Possession of potential "burglar tools" can be used as evidence against you if you are found in incriminating circumstances. An example of a state law to this effect can be found in the Virginia State Code: Section 18.2-94 —Possession of burglarious tools, etc.—"If any person have in his possession any tools, implements or outfit, with intent to commit burglary, robbery, or larceny, upon conviction thereof shall be guilty of a Class 5 felony."

Note that the prosecution has to prove intent. However the law continues, "The possession of such burglarious tools, implements or outfit by any person other than a licensed dealer (these would include, Locksmith, Repossessor, Lock Manufacturer or Dealer, Motor Vehicle Manufacturer or Dealer) shall be prima facie evidence of an intent to commit burglary, robbery, or larceny." This means that the possessor of such tools may have a bit of a problem, to say the least, in trying to convince a jury that this "prima facie evidence" is misleading.

Since you are going to be a locksmith you won't have any problems with possession of these tools as long as you can avoid incriminating circumstances. You must do this at all costs. A criminal act that would have been a misdemeanor becomes a felony if you are a locksmith. Also you would be out of business instantly and you would have many locksmiths very angry that you disgraced and damaged their trade. One bad apple in this trade does affect us all. The ramifications are almost endless.

Relevant laws can be dealing with burglary, motor vehicles, locksmith regulation, and so forth. A law in the state where we live can be completely different from the laws in the area where you live. It is important that you find out what the laws are for your area and determine the applicability to your circumstances (e.g., locksmith, full- or part-time, repo man, etc.).

Find Out the Laws

It is important that you find out what the laws are for your area. You can probably find this information by calling your local district attorney's office or other law offices in your area. Begin by calling your state attorney's office and ask for a copy of the rules and regulations for operating a locksmith business.

We will emphasize the importance of this again. Since laws are different from city to city and from state to state we cannot be responsible for the accuracy of the information in this section as it relates to you or your business.

WARNING: You must find out what the laws are **for your area.**

Trade Associations

SECTION OUTLINE

SECTION 14

Trade Associations

Introduction

What is a Trade Association? Locksmith trade associations are groups of individual locksmiths who have come together with a desire to make the locksmith industry better. Better for the community. Better for the locksmiths. Better for the industry. Members cooperate in order to educate themselves, to make more money, and to make the trade more fun.

The list of associations we have provided for you here is not complete. New groups are being formed daily. Before you contact any of the groups on this list, check with the locksmiths in your area to find out if there is a local group in your town. Also find out which associations the locksmiths in your area belong to and ask questions! You can learn more about trade associations in the section on **Training and Education.**

 **Trade Associations**

NATIONAL ASSOCIATIONS

American Lock Collectors Association
36076 Grennada
Livonia, MI 48154
Phone: 313-522-0920

American Society for Industrial Security
1655 N. Ft. Myer Dr., Ste. 1200
Arlington, VA 22209
Phone: 703-522-5800

Associated Locksmiths' of America, Inc.
3003 Live Oak St.
Dallas, TX 75204
Phone: 214-872-1701

Institutional Locksmiths' Association
P.O. Box 450
Falls Church, VA 22040-0450

Lock Museum of America
P. O. Box 104
Terryville, CT 06786-0114
Phone: 860-589-6359

National Locksmith Suppliers Association
1900 Arch St.
Philadelphia, PA 19103
Phone: 215-564-3484

Safe and Vault Technician's Association
3003 Live Oak St.
Dallas, TX 75204-6186
Phone: 214-827-7233

INTERNATIONAL ASSOCIATIONS

Australia

Locksmiths Guild of Australia, Inc.
P.O. Box 82
Ramsgate, NSW 2217
Australia
Phone: (61) 2-529-7559

Belgium

Belgian Locksmith Federation—BLF
Brussels Chamber of Commerce & Industry
Ave. Louise 500, B1050
Brussels, Belgium
Phone: (32) 2-648-50-02

Denmark

Danish Locksmiths Association—DLF
Allegade 23, DK-2000
Frederiksberg, Denmark
Phone: (45) 38 34 89 07

England

Master Locksmiths Association
Units 4/5 The Business Park
Woodford Halse, Davenry
Northants NN11 3PZ England
Phone: (44) 01327 26225

Finland

Finish Locksmiths Association—FLF
Fredrikinkatu 57, SF-00100
Helsinki, Finland
Phone: (358) 0 694 7044

Germany

European Locksmiths Association—ELF
Muehlentorstrasse 17, D-49808
Lingen (Ems), Germany
Phone: +49 591 3399

Interkey
Muehlentorstrasse 17, D-49808
Lingen, Germany
Phone: (49) 591 51079

Holland

Dutch Locksmith Organization—NSSG
P.O. Box 5254, NL-3008 Ag
Rotterdam, Holland
Phone: (31) 76 14 25 61

Hong Kong

Hong Kong Chapter of ALOA
901 Canton Rd., G/Fl
Mong Kok Hong, Hong Kong

Korea

Korea Chapter of ALOA
1250-7 Bi San 7 Dong
Seo Gu Tae Gu, Korea

Norway

Norwegian Locksmiths Association—NL
Postsboks 70, N-1406
Hebekk, Norway
Phone: (47) 64 87 18 56

Spain

**APECS—Asociacion de Profesionales de Espana
en Cerrajeria y Seguidad Arragueta**
3 bajo P.O. Box 526
E-20600 Eibar, Spain
Phone: (34) 9-43-701391

Sweden

Swedish Master Locksmiths Association—SLR
Box 284, S-127 25
Skarholmen, Sweden
Phone: (46) 8 740 59 55

REGIONAL ASSOCIATIONS

Alabama

Alabama Locksmith Association
1230 Adell St.
Prattville, AL 36066
Phone: 334-271-2223

Alaska

Alaska Locksmith Association
Box 72895
Fairbanks, AK 99707

Arizona

Grand Canyon Chapter of AOLA
13016 S. 131st St.
Gilbert, AZ 85233
Phone: 620-786-8182

Professional Associated Locksmiths of Arizona, Inc.
29 W. Thomas Rd., Ste. A
Phoenix, AZ 85013
Phone: 602-263-9777

Southern Arizona Locksmiths Association
P.O. Box 40001
Tucson, AZ 85717

California

California Locksmiths Association
1240 N. Jefferson St., Ste. G
Anaheim, CA 92807
Phone: 714-632-6800 or
1-800-SOS-LOCK

L.A./Orange Counties Chapter of ALOA
P.O. Box 697
Alhambra, CA 91802-0697
Phone: 310-869-2558

San Diego Chapter of ALOA
1049 Island Ave.
San Diego, CA 92101

Security Locksmiths Association
1401 W. Temple St.
Los Angeles, CA 90026
Phone: 213-413-1121

Canada

The Locksmiths Association of Ontario
2220 Midland Ave., Unit 206
Scarborough, ON M1P 3E6
Canada
Phone: 416-321-2219

British Columbia Locksmiths Association
637 Pine St.
Qualicam Beach, BC V9K 1J1
Canada

Master Locksmiths of Quebec, Inc.
C.P. 65076, Place Longueuil
Longueuil, PQ J4K 5E6
Canada
Phone: 514-463-2759

Professional Locksmiths Association of Alberta
Box 68060 Bonnie Doon P.O.
Edmonton, AB, T6C 4N6 Canada
Phone: 403-948-9997

Colorado

Central and Southern Colorado Locksmith Association
P.O. Box 392
Colorado Springs, CO 80901

Colorado Front Range Chapter of ALOA
2603 Pearl St.
Boulder, CO 80302

Rocky Mountain Locksmith Association
P.O. Box 2751
Denver, CO 80201
Phone: 303-347-8770

Western Slope Locksmith Association
306 N. 12th St.
Gunnison, CO 81230
Phone: 303-641-3940

Connecticut

Locksmith Association of Connecticut, Inc.
41 Garfield Ave., 3rd Floor
Bridgeport, CT 06606

District of Columbia

Locksmiths' Association of Washington DC Area Inc.
P.O. Box 501
Garrett Park, MD 20896-0501

Florida

Central Florida Locksmith Association
2016 Woodland St.
New Smyrna Beach, FL 32168
Phone: 904-428-7233

Florida–Alabama Locksmith Association
6 Comet St. SW
Ft. Walton Beach, FL 32548
Phone: 904-243-6919

Florida West Coast Chapter of ALOA
4410 McElroy Ave.
Tampa, FL 33611
Phone: 813-831-4433

Northwest Florida Locksmith Association
93 Monahan Dr.
Fort Walton Beach, FL 32547
Phone: 904-862-8221

South Florida Chapter of ALOA
1710 NE Miami Gardens Dr.
Miami, FL 33179
Phone: 305-944-0469

South Florida Locksmiths Association
P.O. Box 14322
Fort Lauderdale, FL 33302

State of Florida Board of Locksmiths Inc.
P.O. Box 215
Sanford, FL 32772-0215
Phone: 407-322-4757

Georgia

Dixie Locksmith Association Inc.
129 Sunset Lane
Albany, GA 31705
Phone: 912-888-7233

Georgia Chapter of ALOA
P.O. Box 48088
Atlanta, GA 30362

Hawaii

Hawaii Chapter of ALOA
95-432 Kamahana Place
Mililani, HI 96789

Illinois

The Greater Chicago Locksmiths Association
6216 W. North Ave.
Chicago, IL 60639
Phone: 773-622-0944

Illinois–Indiana Locksmith Association
P.O. Box 368935
Chicago, IL 60636
Phone: 708-422-4808

Illinois Locksmith Association
1120 East Park
Taylorville, IL 62568
Phone: 217-824-8696

Northern Illinois Locksmith Association
38482 N. Sheridan Rd.
Waukegan, IL 60087

Indiana

Central Indiana Chapter of ALOA
1237 Wabash Ave.
Terre Haute, IN 47807

Northern Indiana Chapter of ALOA
122 N. Orchard St.
Kendallville, IN 46755

Southern Indiana Chapter of ALOA
P.O. Box 133
Huntingburg, IN 47542

Iowa

Iowa Locksmith Association
1000 Hillcrest
Sac City, IA 50583
Phone: 712-662-4455

Kansas

Great Plains Locksmith Association
1715 Ave. A
Dodge City, KS 67801
Phone: 316-624-6270

Kentucky

Central States Locksmith Association
1112 Winchester Rd.
Lexington, KY 40505
Phone: 606-253-4811

KYANA Chapter of ALOA
510 E. Parrish Ave.
Owensboro, KY 42303-3125

Louisiana

ARK–LA TEX Locksmith Association
P.O. Box 44337
Shreveport, LA 71134-4437
Phone: 318-222-8378

Louisiana–Mississippi Locksmith Association, Inc.
101 Carter St.
Vidalia, LA 71373
Phone: 318-336-5288

Maine

Pine Tree State Locksmith Association, Inc.
79 Chestnut St.
Lewiston, ME 04240

Maryland

Chesapeake Chapter of ALOA
P.O. Box 12398
Baltimore, MD 21281
Phone: 301-236-4115

Institutional Locksmiths Association
3866 Stoneybrook Rd.
White Plains, MD 20695

Maryland Locksmiths Association
P.O. Box 12398
Baltimore, MD 21281-2398
Phone: 301-236-4115

Massachusetts

Institutional Locksmiths Association
P.O. Box 4097
Dedham, MA 02027
Phone: 617-789-3274

Massachusetts Chapter of ALOA
200 Weir St.
Taunton, MA 02780

Michigan

Locksmith Security Association
19559 14 Mile Rd.
Clinton Twp., MI 48035
Phone: 810-791-5416

Michigan Master Locksmiths Association
26049 W. Warren
Dearborn Heights, MI 48127

Minnesota

Minnesota Chapter of ALOA
41251 Fisher Ave., NE
Prior Lake, MN 55372
Phone: 612-894-1700

Missouri

Gateway Locksmith Association
P.O. Box 746
Bridgeton, MO 63044-0746
Phone: 314-351-7252

Missouri–Kansas Locksmith Association
P.O. Box 11263
Kansas City, MO 64119
Phone: 913-562-3494

Montana

Montana Chapter of ALOA
1000 S. Main
Butte, MT 59701

Nebraska

Nebraska Locksmiths Association
343 W. 2nd St.
Hastings, NE 68901

Nevada

Nevada Professional Locksmiths Association, Inc.
875 S. Boulder Hwy.
Henderson, NV 89105
Phone: 702-566-4363

Sierra Nevada Chapter of ALOA
P.O. Box 115
Reno, NV 89504

New Hampshire

New Hampshire Locksmiths Association, Inc.
124 Union Rd.
Strattham, NH 03885
Phone: 603-772-8524

New Jersey

Garden State Chapter of ALOA
106 Ridgedale Ave.
Morristown, NJ 07960

Master Locksmith Association of NJ
P.O. Box 2441
Morristown, NJ 07962-2441
Phone: 201-538-1588

New Jersey Locksmiths Association
P. O. Box 941
Piscataway, NJ 08854
Phone: 908-780-1550

North Jersey Master Locksmiths Association
99 Park Ave.
Park Ridge, NJ 07656
Phone: 201-391-0081

South Jersey Locksmiths Association
P.O. Box 488
Morton Ave.
Rosenhayn, NJ 08532
Phone: 609-455-2161

New Mexico

New Mexico Chapter of ALOA
917 2nd St., NW
Albuquerque, NM 87102

New Mexico Locksmith Association
4200 Wyoming NE
Albuquerque, NM 87111
Phone: 505-293-6552

Sun Belt Chapter of ALOA
121 Wyatt, Ste. 15
Las Cruces, NM 88005
Phone: 505-524-1000

New York

Adirondack Hudson Master Locksmith Association
P.O. Box 6310
Albany, NY 12206
Phone: 518-462-5467

Central New York Locksmiths Association
15 Garden Ave
Binghamton, NY 13904
Phone: 607-722-9514

Nassau–Suffolk Master Locksmith Association, Inc.
112 Broadway
Amityville, NY 11701
Phone: 516-691-5837

North Carolina

North Carolina Locksmiths Association, Inc.
3232 Maplewood Ave.
Mebane, NC 27302

North Dakota

North Dakota Chapter of ALOA
Rt 1 Box 67
Devil's Lake, ND 58301-3907
Phone: 701-662-5625

Northern Prairie Locksmith Association
2610 Main Ave.
Fargo, ND 58103
Phone: 701-293-7138

Ohio

Ohio North Coast Chapter of ALOA
2 Horseshoe Dr.
Monroeville, OH 44847
Phone: 419-465-4153

Ohio Valley Chapter of ALOA
9178 Reading Rd.
Reading, OH 45215

Penn–Ohio Locksmiths Association, Inc.
P.O. Box 284
Burton, OH 44021
Phone: 216-297-6466

Professional Locksmith Guild
P.O. Box 7217
Milford, OH 45150
Phone: 513-248-9659

Oklahoma

Oklahoma Master Locksmith Association
10001 D.J.I. Drive
Jones, OK 73049
Phone: 405-769-4966

Oregon

Oregon Association of Professional Locksmiths
2550 Santiam Hwy. SW
Albany, OR 97321
Phone: 503-928-9417

Pacific Locksmith Association
P.O. Box 68
Dallas, OR 97338
Phone: 503-623-5315

Pennsylvania

Central Pennsylvania Locksmith Association
1937 Princeton Ave.
Camp Hill, PA 17011
Phone: 717-763-9486

Institutional Locksmiths Association
P.O. Box 24772
Philadelphia, PA 19111

Keystone State Chapter of ALOA
P.O. Box 261
Eagleville, PA 19408-0261

The Greater Philadelphia Locksmiths Association
P.O. Box 11101
Philadelphia, PA 19136

Penn–Ohio Locksmiths Association, Inc.
P.O. Box 284
Burton, OH 44021
Phone: 216-297-6466

Western Pennsylvania Locksmith Association
1009 4th Ave.
Coraopolis, PA 15108
Phone: 412-262-7221

Rhode Island

Locksmith Association of Rhode Island
P.O. Box 9181
Providence, RI 02940-9181

South Carolina

Institutional Locksmith Association
2435 Forest Dr.
Columbia, SC 29204
Phone: 803-256-5495

South Carolina Locksmith Association
P.O. Box 264
Camden, SC 29020
Phone: 803-432-8099

South Dakota

South Dakota Chapter of ALOA
402 S. Minnesota Ave.
Sioux Falls, SD 57102

Tennessee

East Tennessee Locksmiths Association
1566 Lawrence St.
Kingsport, TN 37665
Phone: 423-247-7788

Smokey Mountain Chapter of ALOA
P.O. Box 18187
Knoxville, TN 37928-2187

West Tennessee Chapter of ALOA
1643 Barlett Rd.
Memphis, TN 38134

Texas

Greater Dallas Locksmith Association
8901 Chancellor Rd.
Dallas, TX 75247
Phone: 214-631-1566

Greater Houston Locksmiths Association
10922 Twigg
Houston, TX 77089
Phone: 713-946-7091

Institutional Locksmiths Association
TR 1 Box 776
Krum, TX 76259

Locksmith Association of San Antonio
P.O. Box 21156
San Antonio, TX 78221
Phone: 210-923-4381

Metroplex Locksmith Association
2113 Cliffside Dr.
Fort Worth, TX 76134

The Texas Locksmiths Association
P.O. Box 217
Lake Jackson, TX 77566
Phone: 409-297-2413

Utah

Beehive State Locksmith Association
P.O. Box 65823
Salt Lake City, UT 84115
Phone: 801-266-5083

Vermont

Green Mountain Locksmiths Association
P.O. Box 8246
Essex, VT 05451
Phone: 802-878-4694

Virginia

Institutional Locksmiths Association
429 Broad St.
Portsmouth, VA 23707
Phone: 804-393-6690

Tidewater Locksmiths Association, Inc.
P.O. Box 68244
Virginia Beach, VA 23471

Virginia Locksmiths Association, Inc.
7418 Sudly Rd.
Manassas, VA 22110

Washington

Columbia Basin Locksmith Association
1612 E. Edison
Sunnyside, WA 98944
Phone: 509-839-5260

Northwest Locksmith Association
P.O. Box 75049
Seattle, WA 98944
Phone: 206-793-1276

Western Washington Chapter of ALOA
121 E. Main Ave.
Puyallup, WA 98372

Wisconsin

Fox Valley Chapter of ALOA
1425 N. Richmond St.
Appleton, WI 54911
Phone: 414-731-5400

Wyoming

Wyoming Locksmith Association
Spearfish, SD 57783
Phone: 605-642-4542

Schools

SECTION OUTLINE

SECTION 15

Schools

Introduction

We have given you a list of schools in this Manual just in case you are interested in continuing your education. We are unable to give costs for instruction at each school, but you can expect to pay between $2,000 to $10,000 and higher for classroom instruction. Correspondence schools are much cheaper. Costs vary greatly from school to school and some schools prefer to customize classes to meet your needs. This is much more cost effective for you.

You may be required to take a test (for a fee) before you enroll for classroom instruction to see where your weaknesses are. You'll receive counseling as to which areas need improvement. They seem to want to make sure you do know what you think you already know before training you in other areas. Look over the chart we have provided for you in this section. Some courses are correspondence courses and some require classroom instruction. We have also included estimated completion times for some of the courses.

Abram Friedman Occupational Center
1646 S. Olive St.
Los Angeles, CA 90015
Phone: 213-742-7657

Academy of Locksmithing
2220 Midland Ave., Unit 106
Scarborough, ON M1P 3E6
Canada
Phone: 416-321-2220

Acme School Locksmithing Div.
11350 S. Harlem
Worth, IL 60482
Phone: 708-361-3750

American Locksmith Institute of Nevada
875 S. Boulder Hwy.
Henderson, NV 89015
Phone: 702-565-8811

California Institute of Locksmithing
14721 Oxnard St.
Van Nuys, CA 91411
Phone: 818-994-7426

Charles Stuart School
1420 Kings Hwy., 2nd Floor
Brooklyn, NY 11229

Colorado Locksmith College
4991 W. 80th Ave.
Westminster, CO 80030
Phone: 303-427-7773

Commercial Technical Institute
116 Fairfield Rd.
Fairfield, NJ 07004
Phone: 201-575-5225 or
1-800-526-0890

Foley–Belsaw Institute
6301 Equitable Rd.
Kansas City, MO 64120
Phone: 1-800-821-3452

Golden Gate School of Locksmithing
3722 San Pablo Ave.
Oakland, CA 94608
Phone: 510-654-2677

Granton Institute of Technology
263 Adelaide St., West
Toronto, ON M5H 1Y3
Canada
Phone: 416-977-3929

Lock and Safe Institute of Technology, Inc.
1650 N. Federal Hwy.
Pompano Beach, FL 33062
Phone: 305-785-0444 or
1-800-457-LOCK

Lockmasters, Inc.
5085 Danville Rd.
Nicholasville, KY 40356
Phone: 606-885-7093 or
1-800-654-0637

Locksmith Business Management School
P.O. Box 8525
Emeryville, CA 94662
Phone: 510-654-2677

Locksmith School
51 Beverly Hills Dr.
Toronto, ON M3L 1A2
Canada
Phone: 416-960-9999

Locksmith School, Inc.
3901 S. Meridian St.
Indianapolis, IN 46217
Phone: 317-632-3979

Locksmithing Institute
116 Fairfield Rd.
Fairfield, NJ 07004
Phone: 201-575-5225 or
1-800-526-0890

Master Locksmith Training Courses
Units 4/5 The Business Park
Woodford Halse, Daventry
Northants NN11 3P2
England
Phone: (44) 01327 262255

Messick Vo-Technical Center
703 South Greer
Memphis, TN 38111
Phone: 901-325-4840

North Bennet St. School
39 N. Bennet St.
Boston, MA 02113-1998
Phone: 617-227-0155

Northern Metro. College of TAFE
Attn. Mr. Max Cherry
Waterdale Rd. and Bell St.
Heidelberg, Vic 3081
Australia
Phone: (61) 3 9242-8687

NRI School of Locksmithing
4401 Connecticut Ave., N. W.
Washington, DC 20008
Phone: 202-244-1600

Pine Technical College
1000 Fourth St.
Pine City, MN 55063
Phone: 1-800-521-7463

Prince George's Community College
301 Largo Rd., Room K-205
Largo, MD 20772
Phone: 301-322-0871

San Francisco Lock School
4002 Irving St.
San Francisco, CA 94122
Phone: 415-566-5545

School of Lock Technology
302 W. Katella Ave.
Orange County, CA 92867
Phone: 714-633-1366

Security Education Plus
400-B Etter Dr.
P.O. Box 497
Nicholasville, KY 40356
Phone: 606-887-6027

Security Systems Management Schools
116 Fairfield Rd.
Fairfield, NJ 07004
Phone: 201-575-5225 or
1-800-526-0890

Southern Locksmith Training Institute
1387 Airline Dr.
Bossier City, LA 71112
Phone: 318-227-9458

Stotts Correspondence College
140 Flinders St.
Melbourne, Vic. 3000
Australia

Sydney Institute of Technology
Locksmithing Section Bldg. P
Thomas St., Ultimo, NSW 2007
Australia
Phone: (61) 02-9217 3449

Universal School of Master Locksmithing
3201 Fulton Ave.
Sacramento, CA 95821
Phone: 916-482-4216

Valley Technical Institute
5408 N. Blackstone
Fresno, CA 93710
Phone: 209-436-3814

Vormingsinstituut Voor KMO
Spoorwegstraat, 14
Brugge B8200
Belgium
Phone: (32) 050-383753

Locksmithing Schools	Auto Locks	Master Keying	Lock Servicing	Safe Work	Electronic Security	Training Facility	Completion Time	Financing Provided
Academy of Locksmithing	Yes	Yes	Yes	Yes	No	Campus	Varies	Yes
Acme School of Locksmithing	Yes	Yes	Yes	Yes	No	Campus	60 hrs.	No
American Locksmith Institute of Nevada	Yes	Yes	Yes	Yes	No	Campus	600 hrs.	No
California Institute of Locksmithing	Yes	Yes	Yes	Yes	Yes	Campus	350 hrs.	Yes
Charles Stuart School	Yes	Yes	Yes	Yes	Yes	Campus	900 hrs.	Yes
Colorado Locksmith College	Yes	Yes	Yes	Yes	No	Campus	200 hrs.	Yes
Commercial Technology Institute	Yes	No	Yes	No	No	Correspondence	Varies	Yes
Foley–Belsaw Institute	Yes	Yes	Yes	Yes	Yes	Correspondence	Varies	Yes
Golden Gate School of Locksmithing	Yes	Yes	Yes	Yes	Yes	Campus or Correspondence	340 hrs.	Yes
Granton Institute of Technology	Yes	Yes	Yes	No	No	Correspondence	Varies	Yes
Lockmasters Inc.	No	Yes	Yes	Yes	Yes	Correspondence	Varies	No
Locksmithing Institute	Yes	Yes	Yes	Yes	No	Correspondence	Varies	Yes
Locksmith School	Yes	Yes	Yes	No	No	Campus	6 wks.	No
Locksmith School Inc.	Yes	Yes	Yes	Yes	No	Campus	60 hrs.	Yes
NRI School of Locksmithing	Yes	Yes	Yes	Yes	Yes	Correspondence	Varies	Yes
North Bennet St. School	Yes	Yes	Yes	Yes	No	Campus	36 wks.	Yes
Pine Technical College	Yes	Yes	Yes	Yes	Yes	Campus	9 mo.	Yes
Prince George's College	Yes	No	Yes	Yes	Yes	Campus	150 hrs.	No
Quintilian Institute	No	Yes	Yes	Yes	No	Campus	5 days	No
San Francisco Lock School	Yes	Yes	Yes	Yes	Yes	Campus	240 hrs.	No
School of Lock Technology	Yes	Yes	Yes	Yes	No	Campus	12–20 wks.	No
Security Education Plus	No	No	No	Yes	Yes	Seminars	1–2 days	Yes
Universal School of Master Locksmithing	Yes	Yes	Yes	Yes	No	Campus or Correspondence	120 hrs.	No
Valley Technical Institute	Yes	Yes	Yes	No	No	Campus	360 hrs.	No
Master Locksmiths Association	Yes	Yes	Yes	Yes	Yes	Campus	Varies	No
Northern Metro College	Yes	Yes	Yes	Yes	Yes	Campus	Up to 920 hrs.	Yes
Stott's Correspondence College	Yes	Yes	Yes	No	Yes	Correspondence	Varies	No
Sydney Institute of Technology	Yes	Yes	Yes	Yes	Yes	Campus	Up to 864 hrs.	No

Suppliers

SECTION OUTLINE

Suppliers (Alphabetical List) includes:

    Company Name

    Address

    Phone Number

    800 Number

Suppliers

Introduction

The following list of suppliers is not complete. We have chosen only the suppliers that supply both locks and locksmith tools. There are many other suppliers not listed here that would be classified as a locksmith supplier but are actually in other fields such as alarm systems or physical security such as burglar bars or iron works. You will want to choose a supplier as close to you as possible for quickest delivery times. Generally, the differences between the suppliers, other than service and cost, will be minimal.

Always shop for price; you will find different suppliers charging various prices for the same product. Some will not charge for shipping (most do however), but the difference in price from one supplier to another can be quite significant.

Ask each supplier you deal with if they will give you a discount if you promise to deal exclusively with them. Some will do this on a percentage basis and will not mind if, when they are out of stock on something you need, you order from another supplier.

The personalities, or the feeling you get from each supplier about their personal service, will vary from company to company. If you do not like the company you've chosen, simply stop doing business with them. You have many suppliers to choose from. Some will try to help you find a product at another company or special order it for you if they do not have it. Some will not—don't use them!

 REMEMBER: Your supplier should have the attitude that they are working for you! After all, you are the reason they are in business.

A good supplier can be an asset to your business. Besides getting you the products you need they also occasionally hold seminars with lock manufacturers on hand to answer questions and give classes on their products. You should be able to call your supplier just to ask questions. Have them send you flyers on products you're not familiar with. A good supplier is another avenue to continued education.

 **Suppliers**

A.A. Security Supply Dist.
85 Alpha Park
Cleveland, OH 44143-2228
Phone: 216-442-0560

Accredited Lock Supply Co.
1161 Paterson Plank Rd.
Secaucus, NJ 07094
Phone: 201-865-5015
1-800-652-2835

Ace Lock and Security Supply
565 Rahway Ave.
Union, NJ 07083
Phone: 908-688-7666
1-800-524-1224

Adams Lock and Safe Co.
130 Hall Street
Concord, NH 03301
Phone: 603-224-1652
1-800-635-2228

ADI
180 Michael Dr.
Syosset, NY 11791
Phone: 516-921-6700
1-800-233-6261

Aetna Hardware
5909 W. Lisbon Ave.
Milwaukee, WI 52310
Phone: 414-445-1800
1-800-242-5577

Akron Hardware Consultants
2579 S. Arlington Rd.
Akron, OH 44319
Phone: 216-644-7167
1-800-321-9602

Akron Hardware Consultants
2877 N. Nevada, Ste. 120
Chandler, AZ 85225
Phone: 1-800-457-9378

American Security Distribution
4411 E. La Palma Ave.
Anaheim, CA. 92807
Phone: 714-996-0791
1-800-844-8545
Other Branches:
Long Island, NY . . 1-800-299-5686
Kansas City, KS . . . 1-800-888-5625
Dallas, TX 1-800-292-0115
Denver, CO 1-800-346-5549
Phoenix, AZ 1-800-826-6968
Seattle, WA 1-800-356-5661
Chicago, IL 1-800-621-0177
Orlando, FL 1-800-681-9947
Fremont, CA 1-800-227-1142

Anaheim, CA 1-800-844-8545
San Diego, CA 1-800-231-8435
Atlanta, GA 1-800-952-4452
Houston, TX 1-800-435-8105
Baltimore, MD 1-800-358-6399
Boston, MA 1-800-291-2797

Amos Lock & Key Supply
2711 Cullen St.
Fort Worth, TX 76107
Phone: 817-335-5397
1-800-772-7432

Anderson Lock Co. Ltd.
850 Oakton St.
Des Plaines, IL 60018
Phone: 847-296-1157
1-800-323-5625

Apex Security Hardware
1201 36th St.
Brooklyn, NY
Phone: 718-438-2215
1-800-232-1117

Aristo Sales Co.
24–27 Jackson Ave.
Long Island City, NY 11101
Phone: 718-361-1040
1-800-221-1322

A.T. Jones & Son, Inc.
1456 Broadway
Detroit, MI 48226
Phone: 313-965-1455

Auto Lock and Assoc.
10997 Live Oak Lane
Adelanto, CA 92301
Phone: 1-800-582-2523

Automatic Access Control
8200 Springwood Dr., #230
Irving, TX 75063
Phone: 214-401-1400
1-800-972-8866

Bell's Security Sales, Inc.
411 Bloomfield Ave.
Bloomfield, NJ 07003
Phone: 201-743-3709
1-800-772-2266

Berg Wholesale, Inc.
P.O. Box 29169
Portland, OR 97229
Phone: 503-220-0000
1-800-243-8887

Blaydes Industries, Inc.
2335 18th Street, NE
Washington, DC 20018
Phone: 202-832-7100
1-800-424-2650

Boston Lock & Safe Co.
30 Lincoln St.
Boston, MA 02135
Phone: 617-787-3400
1-800-252-5757

Cal-Royal Products, Inc.
2110 Tubeway Ave.
Commerce, CA 90040
Phone: 213-888-6601
1-800-876-9258

Caola & Co.
2 Crossroads Dr.
Trenton, NJ 08619
Phone: 609-890-7331
1-800-257-9469

Capitol Lock Co. Inc.
9815 Rhode Island Ave.
College Park, MD 20740-1423
Phone: 301-513-9500

C.C. Craig Company Ltd.
1500 King Edwards St.
Winnipeg, MB
R3H 0R5, Canada
Phone: 204-633-9192
1-800-665-7323

**Central Lock & Hardware
  Supply Co.**
95 NW 166th St.
Miami, FL 33169-6048
Phone: 305-947-4853
1-800-677-4549

City Lock Supply
2900-C Valmont Rd.
Boulder, CO 80301
Phone: 303-444-4407

Clark Security Products
4775 View Ridge Ave.
San Diego, CA 92123
Phone: 619-505-1950
1-800-854-2088

Clark Security Products
135 West 2950 South
Salt Lake City, UT 84115
Phone: 801-487-3227
1-800-453-6430

Clark Security Products
4720 Boston Way, Ste. K
Lanham, MD 20706-4310
Phone: 301-731-4100
1-800-578-5625

Clark Security Products
6260 N. Washington St., Unit 25
Denver, CO 80216-1112
Phone: 303-288-9200
1-800-282-5625

Clark Security Products
211 S. 28th St.
Phoenix, AZ 85034
Phone: 602-275-4431
1-800-775-5625

Clark Security Products
3701 Seaport Blvd.
W. Sacramento, CA 95691-3558
Phone: 916-372-6630
1-800-245-3003

Clark Security Products
2760 4th Ave. S.
Seattle, WA 98134
Phone: 206-467-3000
1-800-942-5275

Clark Security Products
51 Faith Ave.
Auburn, MA 01501
Phone: 508-832-5370
1-800-746-5625

Clark Security Products
5304 Lindbergh Lane
Bell, CA 90201
Phone: 213-262-2666
1-800-889-5625

Clark Security Products
830 Sivert Dr.
Wood Dale, IL 60191
Phone: 708-350-8500
1-800-755-5625

Clark Security Products
1516 Interstate Dr.
Erlanger, KY 41018
Phone: 606-746-6500
1-800-659-5625

C-L-K Supplies
W. 91 Commerce Dr.
P.O. Box 678
Hayden Lake, ID 83835
Phone: 208-772-4092
1-800-848-6989

Colonial Lock Supply Co., Inc.
7000-G Newington Rd.
P.O. Box 1417
Newington, VA 22122
Phone: 703-550-0770
1-800-732-9117

Commonwealth Lock Co.
1853 Massachusetts Ave.
Cambridge, MA 02140
Phone: 617-876-3301
1-800-442-7009

Cook's Building Specialties, Inc.
2441 Menaul Blvd. NE
P.O. Box 37320
Albuquerque, NM 87176-7320
Phone: 505-833-5701

Craftmaster Hardware Co., Inc.
134 Liberty St.
Hackensack, NJ 07601
Phone: 201-646-9355
1-800-221-3212

Curtis Industries
6140 Parkland Blvd.
Mayfield Heights, OH 44124
Phone: 216-951-2400
1-800-555-5397

Cypress Security Products Ltd.
4101 19th St. NE, Ste. 5A
Calgary, AB T2E 6X8 Canada
Phone: 430-250-1967
1-800-561-1967

D & S Products Inc.
9451 Jackson Rd.
Sacramento, CA 95826
Phone: 916-362-3502
1-800-266-3502

D.G. Maclachlan Ltd.
4050 Grand St.
Burnaby, BC V5C 3N5
Canada
Phone: 604-294-6000
1-800-665-0535

DiMark International Inc.
3117 Liberator St., Unit A
Santa Maria, CA 93455
Phone: 805-922-1182
1-800-235-2435

Dire's Lock Co. Inc.
2201 Broadway
Denver, CO 80205
Phone: 303-294-0179

Doyle Lock Supply Inc.
2211 West River Road North
Minneapolis, MN 55411
Phone: 612-521-6226
1-800-333-6953

D. Silver Hardware Co. Inc.
591 Ferry St.
Newark, NJ 07105
Phone: 201-344-3963
1-800-222-2915

Dugmore & Duncan, Inc.
30 Pond Park Rd.
Hingham, MA 02043
Phone: 617-740-1101
1-800-225-1595

Dugmore & Duncan, Inc.
3629 Reynolds Rd.
Lakeland, FL 33803
Phone: 1-800-232-1595

Dugmore & Duncan, Inc.
9251 Orco Pkwy.
Riverside, CA 92509
Phone: 1-800-325-1595

Edw. Saucedo & Son Co. Inc.
709-711 N. Copia
El Paso, TX 79903-4405
Phone: 915-566-7101
1-800-258-3726

E.L. Reinhardt Co.
3250 Fanum Rd.
Vadnais Heights, MN 55110-5219
Phone: 612-481-0566
1-800-328-1311

Empire Security Supplies
4600 B Nesconset Hwy.
Port Jefferson Station, NY 11776
Phone: 516-928-1919

Ewert Wholesale Hardware, Inc.
4709 W. 120th St.
Alsip, IL 60658
Phone: 708-597-0059
1-800-451-0200

Fairway Supply, Inc.
2631 Lombardy Lane
Dallas, TX 75220
Phone: 214-350-0021
1-800-776-FAIR

Fairway Supply Inc.
4303 Dacoma
Houston, TX 77092
Phone: 713-957-2160
1-800-767-FAIR

Foley–Belsaw Co.
Locksmith Supply Division
6301 Equitable Rd.
Kansas City, MO 64120
Customer Service: 816-483-4200
Orders: 1-800-821-3452

Fradon Lock Co., Inc.
467 Burnet Ave.
Syracuse, NY 13203
Phone: 315-472-6988
1-800-447-0591

Fried Bros., Inc.
467 N. 7th St.
Philadelphia, PA 19123
Phone: 215-627-3205
1-800-523-2924

**Garden State Hardware
Wholesalers**
322–324 W. Front St.
Plainfield, NJ 07060
Phone: 201-753-1343
1-800-544-0616

Gil-Ray Tools Inc.
1306 McGraw St.
Box 801
Bay City, MI 48707
Phone: 517-892-6870

Gracie's Wholesale & Supply Div.
214 E. 2nd St.
Hastings, NE 68901-5217
Phone: 402-462-2737

Great Lakes Lock Distributors
2310 State St.
Erie, PA 16503
Phone: 814-459-0456
1-800-543-8837

**H & H Lock & Security
Wholesalers**
9140 Gaither Rd.
Gaithersburg, MD 20877
Phone: 301-948-1996
1-800-772-9811

Hans Johnson Co.
8901 Chancellor Row
Dallas, TX 75247
Phone: 214-879-1500
1-888-879-1500

Hardware Agencies Ltd.
1220 Dundas St. E.
Toronto, ON M4M 1S3
Canada
Phone: 416-462-1919
1-800-268-6741

Hardware Specialties
Rt. 255 N.
R.D.2 Box 28A
Dubious, PA 15801
Phone: 814-371-8694

Hardware Suppliers of America, Inc.
213 Mill St.
P.O. Box 2208
Winterville, NC 28590
Phone: 919-355-9400
1-800-334-5625

Helen's Keys and Locks
1145 Jude St.
Schenectady, NY 12303-3372
Phone: 518-355-7000

H.E. Mitchell Co.
P.O. Box 14009
Portland, OR 97214
Phone: 503-236-9444
1-800-547-0925
1-800-626-5625

Herbert L. Flake Co.
5005 Gulf Freeway (I-45 South)
Houston, TX 77023
Phone: 713-926-3200
1-800-231-4105

H. S. & S. Wholesale Distributors
12915 W. Eight Mile Rd.
Detroit, MI 48235
Phone: 313-342-6777
1-800-521-2202

IDN-Acme, Inc.
P.O. Drawer 13748
New Orleans, LA 70185
Phone: 504-837-7315
1-800-788-2263
Branches:
Fort Worth, TX 1-800-859-2263
Houston, TX. 1-800-359-2263
San Antonio, TX . . . 1-210-545-3396
Dallas, TX. 1-800-372-2263
Oklahoma City,
   OK 1-405-942-8750
Phoenix, AZ 1-800-525-3131
Denver, CO. 1-800-445-4008

IDN-Armstrong's, Inc.
1440 Dutch Valley Place, NE
Atlanta, GA 30324
Phone: 404-898-8740
1-800-726-3332

**IDN-International Distribution
Network**
Branches:
Atlanta, GA. 404-875-0136
Calgary, AB Canada . . 403-291-4844
Chicago, IL. 708-456-9600
Cincinnati, OH 513-271-8530
Cleveland, OH 216-642-3900
Dallas, TX. 972-664-1240
Davenport, IA. 319-391-8366
Detroit, MI 810-759-3658
Edmonton, AB
   Canada 403-944-0014
Fort Worth, TX. 817-284-5696
Grand Rapids, MI . . . 616-534-1067
Houston, TX. 713-668-0022
Indianapolis, IN 317-635-8100
Jacksonville, FL. 904-387-0663
Milwaukee, WI 414-252-4414
New Orleans, LA . . . 504-837-7315
Norfolk, VA 804-853-0611
Oklahoma, OK 405-942-8750
Orlando, FL 407-297-7722
Ottawa, ON Canada . 613-749-2172

Philadelphia, PA 215-288-5588
Phoenix, AZ. 602-263-5286
Tampa, FL 813-886-8007
Toronto, ON. 416-248-5625
Vancouver, BC
   Canada 604-253-0017

Independent Hardware Inc.
14 S. Front St.
Philadelphia, PA 19106
Phone: 215-925-5306
1-800-346-9464

Infiniti Hardware Supply
2309 Faithful Ave.
Saskatoon, SK S7K 1T9
Canada
Phone: 306-931-0001
1-800-667-6622

Intermountain Lock & Supply Co.
3106 S. Main St.
P.O. Box 65158
Salt Lake City, UT 84165-0158
Phone: 801-486-0079
1-800-453-5386

Intermountain Lock & Supply Co.
2300 W. 2nd Ave., Unit B
Denver, CO 80223
Phone: 303-698-1898
1-800-323-8046

Island Key Supplies
845 Fort St.
Victoria, BC V8W 1H6
Canada
Phone: 604-384-4122

Island Pacific Distributors Inc.
1668 King St.
P.O. Box 22189
Honolulu, HI 96822
Phone: 808-955-1126

Jack Stearman Ltd.
338 W. 6th Ave.
Vancouver, BC V5Y 1K9
Canada
Phone: 604-872-8415
1-800-663-5397

J. Ethier & Sons, Inc.
70 North St.
P.O. Box 921
Fitchburg, MA 01420
Phone: 508-342-0912
1-800-36 ETHIER

J. Nathan Hardware Specialties, Inc.
161 Comfort St.
P.O. Box 115
Rochester, NY 14620
Phone: 716-325-3330
1-800-634-2580

Jo-Van Distributors, Inc.
929 Warden Ave.
Scarborough ON M1L 4C6
Canada
Phone: 416-752-7210

KDL Hardware Supply, Inc.
1621 8th Ave.
P.O. Box 21226
Seattle, WA 98101
Phone: 206-682-7383
1-800-926-7716

Kenco Supply Co.
2531 N. 85th St.
Omaha, NE 68134
Phone: 402-397-8291
1-800-228-2266

Kentucky Lock & Safe
1112 Winchester Rd.
Lexington, KY 40505
Phone: 606-253-4820

Key Products
2555 International St.
Columbus, OH 43228
Phone: 614-876-7704
1-800-457-1019

Key Sales & Supply Co.
9950 Freeland Ave.
Detroit, MI 48227
Phone: 313-931-7720
1-800-445-5397

Key Supply
362 7th St.
San Francisco, CA 94103
Phone: 415-626-2526

Kimko Lock Industry Co. Ltd.
4409 Watermill Ave.
Orlando, FL 32817
Phone: 407-657-4165

Klein Bros. Inc.
1101 W. Broadway
Louisville, KY 40203
Phone: 502-587-6886

Kupferman & Co., T. L.
243-07 131st Ave.
Rosedale, NY 11422
Phone: 718-527-6363

LDM Enterprises
638 Lindero Canyon Rd., Ste. 255
Agoura, CA 91301
Phone: 1-800-451-5950

**Lock & Key Wholesale
Distributors Inc.**
1229 1st Ave. S.
Birmingham, AL 35233
Phone: 205-252-2427

Lockmasters, Inc.
5085 Danville Rd.
Nicholasville, KY 40356
Phone: 606-885-6041
1-800-654-0637

Locks Co.
2050 NE 151st St.
N. Miami, FL 33162
Phone: 305-949-3619
1-800-288-0801

The Locksmith Store, Inc.
1229 E. Algonquin Rd., Ste. E
Arlington Heights, IL 60005
Phone: 847-364-5111

Lockwise Products Inc.
2001 NW 167th St.
Miami, FL 33056
Phone: 305-625-5525
1-800-447-6616

L.V. Sales, Inc.
1831 Hyperion Ave.
Los Angeles, CA 90027
Phone: 213-661-4746
1-800-894-KEYS

Majestic Lock Co., Inc.
194 Daniel St.
Hackensack, NJ 07601
Phone: 201-343-7728
1-800-441-9324

Major Lock Supply
2512 E. Fender Ave., Ste. F
Fullerton, CA 92631-4436
Phone: 714-447-8363
1-800-734-4539

Mayflower Sales Co. Inc.
614 Bergen St.
Brooklyn, NY 11238
Phone: 718-622-8785
1-800-221-2052

Maziuk & Co. Inc.
1251 W. Genesee St.
Syracuse, NY 13204
Phone: 315-474-3959
1-800-777-5945

**McDonald–Dash Locksmith
Supply Inc.**
5767 E. Shelby Dr.
P.O. Box 752506
Memphis, TN 38175-2506
Phone: 901-797-8000
1-800-238-7541

McManus Locksmith Supply Inc.
1309 Central Ave.
P.O. Box 9321
Charlotte, NC 28205
Phone: 704-333-9112
1-800-438-6567

Metro Safe Co., Inc.
2627 E. Mile Rd.
Warren, MI 48091
Phone: 810-755-3570
1-800-541-8050

Mid South Locksmith Supply, Inc.
4176 Getwell
P.O. Box 18468
Memphis, TN 38181
Phone: 901-795-6987
1-800-238-6166

Midwest Wholesale Hardware
5121 Front St.
Kansas City, MO 64120
Phone: 1-800-621-6581

Midwest Wholesale Hardware
997 W. Kenny Blvd., Ste. A18
Orlando, FL 32810
Phone: 1-800-659-8527

Mountain West Alarm Supply Co.
P.O. Box 10780
Phoenix, AZ 85064
Phone: 602-971-1200
1-800-528-6169

Moylan Enterprises Co., Inc.
Moylan Bldg.
260 Soledad Ave.
Agana, Guam 96910
Phone: 011-671-472-6738

M. Shepse Co.
2026 E. Carson St.
Pittsburgh, PA 15203
Phone: 412-381-4900
1-800-666-6007

M. Zion Co. Inc.
17 Murray St.
New York, NY 10007
Phone: 212-349-8677

N & N Distributing
5631 Madison Ave.
Indianapolis, IN 46227
Phone: 317-784-1298

National Safe Co., Inc.
7 Lipson Ct., Ste. 100
E. Newport, NY 11731
Phone: 516-368-2000
1-800-SAFE-640

New York Key Service
1054 N. Western Ave.
Los Angeles, CA 90029
Phone: 213-469-2183

O'Brien Brothers, Inc.
380 Union St.
W. Springfield, MA 01089
Phone: 413-734-7121
1-800-343-0949

Orchard Lock Distributors
30 Edmond St.
Hampton, CT 06517
Phone: 203-865-8106
1-800-233-2146

Page 3 Products, Inc.
565–571 Windsor St.
Hartford, CT 06120
Phone: 203-527-2135
1-800-333-1086

Pasek Corp.
9 W. Third St.
Boston, MA 02127
Phone: 617-269-7110
1-800-628-2822

P. Coudriau & Sons Co. Ltd.
47 Laval St.
Ottawa, ON K1L 7Z7
Canada
Phone: 613-741-7900

Pimlico Key Service, Inc.
5254 Reisterstown Rd.
Baltimore, MD 21215
Phone: 410-367-7400
1-800-638-3815

R & H Wholesale Supply Inc.
1655 Folsom St.
San Francisco, CA 94103
Phone: 415-863-0404
1-800-FOR-LOCK

Robert Skeels & Co.
19216 S. Laurel Park Rd.
Compton, CA 90220
Phone: 310-639-7240
1-800-734-4539

Salz Lock & Safe Co.
4420 Lawehana St., Unit 4
Honolulu, HI 96818-3141
Phone: 808-423-7200

Samaco Hdwe. & Supply Co.
4125 Olive St.
St. Louis, MO 63108
Phone: 314-533-8159

Security Lock Distributors
40 A Street
P.O. Box 815
Needham Heights, MA 02194
Phone: 617-444-1155
1-800-847-5625

Security Plus, Inc.
3612 N. 16th Street
Phoenix, AZ 85016
Phone: 602-234-3883
1-800-426-0200

Security Resources, Inc.
8400 Pittman Ave.
P.O. Box 15532
Pensacola, FL 32514
Phone: 904-476-2799

Semel Goldman Inc.
7 Essex St.
P.O. Box 1041
New York, NY 10002
Phone: 212-674-6401

Sentry Security Fasteners, Inc.
8109 N. University
Peoria, IL 61615-0165
Phone: 309-693-2800

Serrubec Inc.
2073 Chartier Ave.
Montreal, Dorval
PQ H9P1H3 Canada
Phone: 514-631-6791
1-800-361-0243

Shield Supply & Services Ltd.
1391 St. James St., Unit 17
Winnipeg, MB R3H 0Z1
Canada
Phone: 204-774-1921

The Shwayder Co.
2335 E. Lincoln
Birmingham, MI 48008
Phone: 313-645-9511

Silver Sales, Inc.
2701-C West 15th St., Ste. 291
Plano, TX 75075
Phone: 214-867-4545
1-800-258-LOCK

Smallwood Locksmith Supply
1008 N. 18th St.
Kansas City, KS 66102
Phone: 913-371-5787

Smith Bros.
2706 Arctic Ave.
Atlantic City, NJ 08401
Phone: 609-344-0046

So-Cal Lock & Supply
2104-B Wilson Ave.
National City, CA 91950-6555
Phone: 619-474-8847
1-800-521-3551

Southern Lock & Supply Co., Inc.
10910 Endeaver Way
P.O. Box 1980
Pinellas Park, FL 33777
Phone: 813-541-5536
1-800-237-2875

Southern Lock & Supply Co., Inc.
4039 NE 10th Ave.
Oakland Park, FL 33334
Phone: 954-568-9669

Southern Lock & Supply Co., Inc.
7758 NW 72nd Ave.
Miami, FL 33166
Phone: 305-885-5779

Southern Lock & Supply Co., Inc.
6012-E Old Pineville Rd.
Charlotte, NC 28217
Phone: 704-527-6777

Southern Lock & Supply Co., Inc.
3627 Clearview Pkwy.
Atlanta, GA 30340-3969
Phone: 770-455-9233

Space Age Locksmith Supplies, Inc.
138 W. Beresford Ave., Ste. D
Deland, FL 32720
Phone: 904-374-3113
1-800-247-7942

Standard Wholesale Hardware, Inc.
42 Ludlow St.
New York, NY 10002
Phone: 212-353-0450
1-800-543-LOCK

Steadfast Corporation
229 Marginal St.
Chelsea, MA 02150
Phone: 617-889-3400
1-800-DIAL-911

Stearman, Ltd.
338 W. 6th Ave.
Vancouver, BC V5Y 1K9
Canada
Phone: 604-872-8415

Stone & Berg
Wholesale Locksmith Supply Co.
99 Stafford St.
Worcester, MA 01603
Phone: 508-753-3551
1-800-225-7405

Strauss Safe & Lock Co.
1801 2nd Ave.
Des Moines, IA 50314
Phone: 515-288-9571
1-800-532-4107

Taylor Security & Lock Co., Inc.
8585 Atlas Dr.
Gaithersburg, MD 20877
Phone: 301-948-7670
1-800-676-7670

TimeMaster, Inc.
2604 SW 17th
Topeka, KS 66604
Phone: 913-232-8705

Tomco
7657 Winnetka Ave., #195
Winnetka, CA 91306
Phone: 818-883-0342

Tweeds Security Hardware Wholesalers
601 Elm Ave.
Portsmouth, VA 23704
Phone: 757-399-2180
1-800-544-4482

U.S. Industrial Products Corp.
96-12 43rd Ave.
Corona, NY 11368
Phone: 212-335-3300

U.S. Lock Corporation
77 Rodeo Dr.
Brentwood, NY 11717
Phone: 516-243-3000
1-800-925-5000

U.S. Lock Corporation
5351 Ramon Blvd.
Jacksonville, FL 32205
Phone: 904-783-3232
1-800-925-5000

U.S. Lock Corporation
4831 Jennings Lane
Louisville, KY 40218
Phone: 502-966-6958
1-800-925-5000

U.S. Lock Corporation
840 N. 10th St.
Sacramento, CA 95814
Phone: 916-446-7000
1-800-925-5000

Veehoff Supply Co.
908 Lake Ave.
P.O. Box 361
Storm Lake, IA 50588
Phone: 712-732-1836

Wacker Hardware Co.
1025 W. Jackson Blvd.
Chicago, IL 60607
Phone: 312-733-6070

Wespac Corp.
342 Harriet St.
San Francisco, CA 94103-4716
Phone: 415-431-6350

Wholesale 4, Inc.
706 S.E. Grand Ave.
Portland, OR 97214
Phone: 503-238-8605
1-800-547-0921

Wilco Supply
5960 Telegraph Ave.
P.O. Box 3047
Oakland, CA 94609
Phone: 510-652-8522
1-800-745-5450

Williams Key Co., Inc.
2206 Locust St.
St. Louis, MO 63103
Phone: 314-231-2411
1-800-325-1779

World Wide Lock Supply
5950 Kester Ave.
Van Nuys, CA 91411
Phone: 818-781-9999
1-800-729-5444

Zipf Lock Co.
830 Harmon Ave.
Columbus, OH 43223
Phone: 614-228-3507
1-800-848-1577

How to reach us...

TO ORDER:

Call our toll-free order line

1-800-791-8529

or

Fax us at

719-266-4367

TO CONTACT US ON THE WEB

To correspond directly with Dennis Collins,
president of Budget Lock & Key, Inc., please e-mail:

 budkeyman@aol.com

To visit our web page:

 www.budgetlockandkey.com

To order the **SLIMBOW** from our secure server on the internet:

 www.slimjims.com

BUDGET LOCK & KEY INC.

6547 N. Academy, Ste. 532
Colorado Springs, CO 80918

719-598-2135

QUICK START LOCKSMITH TRAINING